Thomas Stolz, Julia Nintemann
Special Onymic Grammar in Typological Perspective

Studia Typologica

Beihefte / Supplements
STUF – Sprachtypologie und Universalienforschung
　　　Language Typology and Universals

Editors
Thomas Stolz, François Jacquesson, Pieter C. Muysken

Editorial Board
Michael Cysouw (München), Ray Fabri (Malta), Steven Roger Fischer (Auckland), Bernhard Hurch (Graz), Bernd Kortmann (Freiburg), Nicole Nau (Poznán), Ignazio Putzu (Cagliari), Stavros Skopeteas (Bielefeld), Johan van der Auwera (Antwerpen), Elisabeth Verhoeven (Berlin), Ljuba Veselinova (Stockholm)

Volume 34

Thomas Stolz, Julia Nintemann

Special Onymic Grammar in Typological Perspective

Cross-Linguistic Data, Recurrent Patterns,
Functional Explanations

DE GRUYTER
MOUTON

ISBN 978-3-11-221549-4
e-ISBN (PDF) 978-3-11-133187-4
e-ISBN (EPUB) 978-3-11-133197-3
ISSN 1617-2957

Library of Congress Control Number: 2023945970

Bibliographic information published by the Deutsche Nationalbibliothek
The Deutsche Nationalbibliothek lists this publication in the Deutsche Nationalbibliografie;
detailed bibliographic data are available on the Internet at http://dnb.dnb.de.

© 2025 Walter de Gruyter GmbH, Berlin/Boston
This volume is text- and page-identical with the hardback published in 2024.
Cover image: Alpha-C/iStock/Thinkstock
Typesetting: Integra Software Services Pvt. Ltd.
Printing and binding: CPI books GmbH, Leck

www.degruyter.com

O fato de não se compreender ou explicar uma coisa não acaba com ela.
(Jorge Amado, *Gabriela, cravo e canela*)

Acknowledgements

This study forms part of the research project *Morphosyntaktische Typologie der Toponyme / Morphosyntactic typology of toponyms (TYPTOP)* (STO 186/27-1; eBer-22-55215) financed by the *Deutsche Forschungsgemeinschaft (DFG)* from 1st of April 2023 until 31st of March 2026. We are grateful to Paula Müller for her kind technical support in the production process of this text. Adriana Bursch kindly let us glimpse at her quantitative data on the Romanian proprial article. Judith Hock was helpful when it came to finding suitable Hungarian prose. We thank Susanne Hackmack and Karl-Heinz Wagner for giving us access to their Swahili corpus data. Rafael Rodríguez-Ponga y Salamanca gave us a Kaqchikel grammar of the 18th century as a present. Kevin Behrens, Nataliya Levkovych, and Maike Vorholt expertly commented on the draft version. We emphasise, however, that everything that is said in this study is solely our own responsibility.

Contents

Acknowledgements —— VII

List of figures —— XI

List of tables —— XIII

Abbreviations —— XV

1   **Introduction** —— 1
1.1   Prelude in the Philippines: Names vs nouns in Iloko —— 2
1.2   The study in a nutshell —— 17

2   **The Grammar of Names vs Special Onymic Grammar** —— 22
2.1   What onomasticians are less interested in —— 22
2.2   The linguistic turn —— 24
2.3   On being special —— 26
2.3.1   A plethora of special grammars: The case of Nungon —— 27
2.3.2   *Special Onymic Grammar* all over the place: Vurës —— 32
2.4   Broadening the horizon —— 37
2.4.1   Beyond German (with glimpses of Palikur and Dargwa) —— 37
2.4.2   Where's typology? —— 42
2.5   Expectations and stipulations —— 47

3   **Empirical bits and pieces** —— 51

4   **Leitmotifs of *Special Anthroponymic Grammar*: Possession** —— 59
4.1   Familiarising ourselves with SAG in the domain of possession —— 60
4.2   Old genitives and new possessives —— 65
4.2.1   Varia Germanica —— 65
4.2.2   A genitive only for names: Faroese —— 75
4.3   Against parsimony: The pronominal possessive in Icelandic —— 82
4.4   Neutralisation: The proprial gender in Romanian —— 92
4.5   Preliminary results: *Special Anthroponymic Grammar* in adnominal possession —— 100

5   **Recurrent themes of *Special Toponymic Grammar*: Spatial relations** —— 106
5.1   Glimpses of Lezgian case-inflection —— 107

5.2	Uralic parallels —— **111**	
5.2.1	Hungarian —— **111**	
5.2.2	Finnish —— **125**	
5.2.3	Estonian —— **133**	
5.2.4	What the Uralic data tell us —— **148**	
5.3	Basque —— **149**	
5.4	Antiquity —— **159**	
5.4.1	Ancient Greek —— **159**	
5.4.2	Latin —— **170**	
5.4.3	Conservatism and structural complexity —— **181**	
5.5	Swahili —— **182**	
5.6	Preliminary results: *Special Toponymic Grammar* in spatial relations —— **195**	

6   **Functional-typological evaluation —— 200**
6.1   A socio-pragmatic long distance parallel: Catalan and Kaqchikel —— **200**
6.2   Predictable and expected —— **207**
6.3   Different ways of expressing the same —— **216**
6.4   Mismatchmaking —— **224**

7   **Conclusions and outlook —— 227**

Primary sources —— 231

References —— 233

Appendix I: List of languages (n = 85) —— 245

Appendix II: Map —— 247

Index of authors —— 249

Index of languages —— 253

Index of subjects —— 255

List of figures

Figure 1	Subclasses of nouns in Vurës —— 33	
Figure 2	Schematic distinction of possessors according to word-class —— 59	
Figure 3	Schematic distinction of possessors according to word-class in Manx —— 60	
Figure 4	Split rule in Amele —— 61	
Figure 5	Split rule in Kryts —— 62	
Figure 6	Split rule in Malagasy —— 64	
Figure 7	Split rule in Japanese dialects —— 65	
Figure 8	Split rule in German I —— 67	
Figure 9	Split rule in German II —— 67	
Figure 10	Split rule in Dutch —— 68	
Figure 11	Split rule in West Frisian —— 70	
Figure 12	Distribution of possessors over constructions in Low German —— 73	
Figure 13	Heavy vs light possessee NPs over construction types in Low German —— 73	
Figure 14	Heavy vs light possessor NPs (ANTHS) over construction types in Low German —— 74	
Figure 15	Opposing tendencies in Low German —— 74	
Figure 16	Split rule in Faroese —— 81	
Figure 17	Competing adnominal possessive constructions in Icelandic —— 85	
Figure 18	Shares of possessee classes over constructions —— 86	
Figure 19	Constructions over possessee classes —— 87	
Figure 20	Split rule in Icelandic —— 92	
Figure 21	Split rule in Romanian —— 99	
Figure 22	Prototypical possessive situation —— 101	
Figure 23	Extended Animacy Hierarchy I —— 102	
Figure 24	Extended Animacy Hierarchy II —— 102	
Figure 25	Schematic distinction of Grounds according to word-class —— 107	
Figure 26	Choice of spatial cases with Lezgian COMMS —— 110	
Figure 27	Choice of spatial cases with Lezgian TOPOS —— 110	
Figure 28	Postpositions governing cases of the superior set —— 116	
Figure 29	Hierarchy of conditions determining the use of spatial cases with TOPOS in Hungarian —— 120	
Figure 30	Split rule for general location in Hungarian —— 124	
Figure 31	Split rule for general location in Finnish —— 132	
Figure 32	Competition of the sets of spatial cases with Estonian TOPOS —— 136	
Figure 33	Split rule for general location in Estonian —— 148	
Figure 34	Continuum of prototypicality / Uralic —— 149	
Figure 35	Distribution of class-markers with Basque nouns —— 152	
Figure 36	Shares of strategies to encode spatial categories in the Odusseias —— 167	
Figure 37	Split rule for general location in Ancient Greek —— 170	
Figure 38	Split rule for general location in Latin —— 181	
Figure 39	Three options of encoding Place in Swahili (preliminary version) —— 183	
Figure 40	Four options of encoding spatial relations in Swahili —— 194	
Figure 41	Main components of spatial situations —— 197	
Figure 42	Implicational pattern for zero-marking —— 210	
Figure 43	Split rule for inalienable possession in Iaai —— 211	

https://doi.org/10.1515/9783111331874-204

Figure 44 Split rule for articles in Gela —— **215**
Figure 45 Split rule for oblique case marking in Halkomelem —— **216**
Figure 46 Split rule for the expression of Place in Dutch —— **218**
Figure 47 Split rule for Place marking in Zulu —— **219**
Figure 48 Split rule for general location in Mussau —— **221**
Figure 49 Split rule for inalienable possession in Mussau —— **223**
Figure 50 Split rule for inalienable possession in Yagaria —— **223**
Figure 51 SOG and noncanonicity —— **224**

List of tables

Table 1	Iloko articles (Rubino 2011: 333) —— 3	
Table 2	Iloko articles (revised) —— 14	
Table 3	Categories relevant for nouny word-classes in Iloko —— 16	
Table 4	Nominal subclasses compared (Sarvasy 2017: 123) —— 28	
Table 5	Properties of subclasses of nouns in Vurēs —— 33	
Table 6	(Im)possible distribution of vocatives over ANTHS and COMMS —— 45	
Table 7	(Im)possible distribution of spatial zero-marking over TOPOS and COMMS —— 46	
Table 8	Five possible types —— 47	
Table 9	Vocative suffixes in Somali —— 57	
Table 10	Paradigms of COMMS and ANTHS compared —— 80	
Table 11	Paradigms of COMMS and ANTHS in Romanian —— 93	
Table 12	Selected spatial cases of Lezgian COMMS and TOPOS —— 108	
Table 13	Selected spatial functions and their expressions with COMMS and TOPOS in Lezgian —— 109	
Table 14	Spatial cases of the interior and superior set of Hungarian TOPOS and COMMS – expression of general location —— 115	
Table 15	Two different TOPO types and pseudo-defectiveness / Hungarian —— 118	
Table 16	The differential use of the interior and superior sets of spatial cases with names of countries in Hungarian —— 122	
Table 17	Distinctive features of classes of nouns in Hungarian —— 123	
Table 18	Spatial cases of the interior and exterior set of Finnish TOPOS and COMMS – expression of general location —— 127	
Table 19	Two different TOPO types / Finnish —— 131	
Table 20	Distinctive features of classes of nouns in Finnish —— 132	
Table 21	Spatial cases of the interior and exterior set of Estonian TOPOS and COMMS – expression of general location —— 141	
Table 22	Distinctive features of classes of nouns in Estonian —— 141	
Table 23	Two different TOPO types / Estonian —— 142	
Table 24	Synonymous word-forms in the paradigm of an Estonian TOPO —— 144	
Table 25	Selected cases of the paradigms of Basque inanimate nouns, animate nouns, and TOPOS —— 153	
Table 26	Continuum of nouny word-classes in Basque —— 159	
Table 27	Paradigms involving spatial clitics in Ancient Greek —— 165	
Table 28	Shapes of the TOPO *Púlos* in the Odusseias —— 168	
Table 29	Distribution of features over word-classes in Ancient Greek —— 169	
Table 30	TOPO split in Latin —— 177	
Table 31	Continuum of shared properties in Latin —— 180	
Table 32	System of ANTH-markers in Kaqchikel —— 203	

Abbreviations

1/2/3	1ˢᵗ/2ⁿᵈ/3ʳᵈ person (or different categories)
Ø	absence of material exponence
ABL	ablative
ABS	absolutive
ACC	accusative
ADESS	adessive
ADJR	adjectiviser
ADV	adverb(ial)
ALL	allative
ANIM	animate
ANTH	anthroponym / anthroponymic marker
ANTH-N	anthronoun
AOR	aorist
APASS	antipassive
APPL	applicative
ART	article
ASSPL	associative plural
AUX	auxiliary
CL	class(ifier)
CL.DOM	classifier for domesticated animals
CMPL	completive
CMPR	comparative
CNJ	conjunction
COM	comitative
COMM	common noun / common noun marker
COND	conditional
CONST	construct (suffix)
CONT	continuous
COP	copula
DAT	dative
DEF	definite
DEL	delative
DELIM	delimiter
DEM	demonstrative
DET	determiner
DIM	diminutive
DIR	directional
DIST	distal
DP	determiner phrase
DS	different subject
DU	dual
DUB	dubitative
DUR	durative
ELA	elative

EMPH	emphatic
ERG	ergative
ESS	essive
EXIST	existential
F	feminine
FOC	focus
FREQ	frequentative
FUT	future
GEN	genitive
GER	gerund
HABIT	habitual
HODPAST	hodiernal past
HON	honorific pronoun
HUM	human
ILL	illative
IMPERF	imperfect
IMPERS	impersonal
IMPTV	imperative
INCH	inchoative
INCMP	incompletive
INDEF	indefinite
INESS	inessive
INF	infinitive
INT	interrogative
INTJ	interjection
INTR	intransitive
INTS	intensive
IPFV	imperfective
IRR	irrealis
LEX	lexeme
LK	linker
LIG	ligature
LOC	locative
M	masculine
MED	medial
MEDV	medial verb form
N	noun
NCONT	non-continuous
NEG	negation
NOM	nominative
NP	noun phrase
NPERS	non-personal article
NPST	non-past
NRRC	non-restrictive relative clause
NSG	non-singular
NT	neuter
OBJ	object

OBL	oblique
O.G.	original glosses
OPT	optative
O.T.	original translation
PART	partitive
PASS	passive
PE	plural exclusive
PERF	perfect
PERL	perlative
PERS	personal article
PFV	perfective
PIV	pivotal marker
PL	plural
PLC	place-name article
PLUP	pluperfect
POR	possessor
PostP	postpositional phrase
POSS	possessive
PP	prepositional phrase
PRED	predicative
PREP	preposition
PRET	preterite
PRO	pronoun
PROG	progressive
PROP	proper name / proper name marker
PROP-N	proper noun
PRS	present
PRX	proximal
PST	past
PTCL	particle
PTCPL	participle
ℜ	(possessive) relation
REC	reciprocal
REL	relativiser / relative
RELL	relationaliser
REMPST	remote past
RFL	reflexive
RLS	realis
RN	relational noun
RRC	restrictive relative clause
RSTR	restrictive
SAG	Special Anthroponymic Grammar
SBJ	subject
SBLTV	sublative
SG	singular
SIM	simultaneous
SOG	Special Onymic Grammar

SS	same subject
STAT	stative
STG	Special Toponymic Grammar
SUBJ	subjunctive (dependent verb form)
SUBORD	subordinator
SUP	superlative
SUPESS	superessive
SW	switch-reference
TEMP	temporal
TERM	terminative
TOPO	toponym / toponymic marker
TOPO-N	toponoun
TRANS	translative
UT	utrum
V	vowel
VIS	visible
VOC	vocative
VR	verbaliser
X, Y, Z	variables for unnamed categories

1 Introduction

The principal goal of this study is to prove empirically that the in-depth investigation of the morphosyntactic behaviour of names across languages promises many new insights into the nature of several domains of linguistic research such as word-class systems, etc. Van Langendonck and Van de Velde (2016: 38) are certainly right when they complain that "grammatical criteria [...] are too often ignored in approaches to names." On the same page, the quoted authors emphasise the necessity of adducing language-specific criteria to determine the grammatical category of names in a given language. We argue that we also need to look beyond the structural facts found in individual languages because the potentially limited scope of the research programme precludes the possibility of discovering linguistically meaningful cross-linguistic similarities between languages which give evidence of "onymische Sondergrammatik", henceforth *Special Onymic Grammar* (= SOG) (Nübling et al. 2015: 64) whose particulars are disclosed in Section 2. SOG is the cover term for all those structural properties of names which set them apart from common nouns. If we read the following few lines from Dixon (2010a: 102) according to which

> [a] proper noun may fill an argument slot in a clause, like a common noun, but it almost always has more limited morphological and syntactic properties. For example, a common noun, but not a proper noun, may inflect for number,

one might want to ask whether SOG is always negative in the sense that names have access usually to only a subset of the grammatical categories to which common nouns are entitled. We doubt that a generalisation of this kind can stand the test as will be shown in this study.

Superficially, the distinctive traits associated with names as opposed to common nouns might look like an assortment of unrelated idiosyncrasies as the data presented in Section 3 seem to suggest. However, typologically-minded language comparison reveals that there also are cross-linguistically recurrent patterns which show that the phenomenology of SOG is by no means entirely arbitrary. To the contrary, SOG keeps coming to the fore in exactly the same domains in language after language so that a relatively high degree of predictability of where in the grammatical system names and common nouns may go different ways can be reached. Sections 3–5 amply document this scenario from several genetically, structurally, and geographically different languages. What is more, the structural dissimilarities between names and common nouns can be explained with reference to functional factors as demonstrated in Section 6. We provide evidence for each and every of our claims throughout this inquiry. In the final Section 7, we summarise the results, draw general conclusions on their basis, and identify future strands of research in the realm of SOG. Before our project is situated in its wider context and methodology, theory, terminology, and database are discussed in Section 1.2, we have a look at a first example of SOG to help the reader to better understand why what we are going to elaborate upon subsequently indeed matters linguistically. Section 1.1 comes in the shape of a case-study in which some of the recurrent phenomena associated with SOG are identified. For a start and only for the purpose of familiarising the reader with the topic, we keep the terminology of the sources we have consulted. Our own terminological choices will be introduced subsequently in Section 1.2. Likewise, Section 1.1 scarcely makes mention of the pertinent literature since subsequent sections will amply discuss the hypotheses put forward by scholars who address phenomena which are of interest for this study.

1.1 Prelude in the Philippines: Names vs nouns in Iloko

In (1), we present three sentential instances of the comparative construction in Iloko (aka Ilocano). In (1a), both comparee (*bosko* 'my voice') and standard (*bosmo* 'your voice') are common nouns. In (1b), two place names (the islands *Oahu* and *Luzon*) are involved in the comparison, whereas (1c) features two personal names of artists as comparee (*Robert Redford*) and standard (*Paul Newman*). The syntactic frame of the constructions is identical for all three sentences. However, *bosko*, *bosmo*, *Oahu*, and *Luzon* are accompanied by *ti* (provisionally glossed X) immediately to their left but the names of the movie stars combine with *ni* (provisionally glossed Y) in the same slot.

(1) Iloko [Austronesian, Northern Luzon] (Espiritu 1984: 199)[1]
    (a) Common noun
        nalá~laing   *ti*   *bos-ko*   *ngem*   *ti*   *bos-mo.*
        ADJR:CMPR~pretty   X   voice-POSS.1SG   but   X   voice-POSS.2SG
        'My voice is prettier than your voice.'

---

[1] In the numbered examples, the object language is identified by glossonym with information about the genetic affiliation added in square brackets. As usual, the source from which a given example has been taken appears in normal brackets. Primary sources are identified in curly brackets and spelled out in a separate section. Those elements in the example which are of special interest for the accompanying discussion are highlighted in bold. If needs be additional marking strategies like underlining are made use of. Square brackets are sometimes used to make syntactic boundaries visible. The glosses follow the *Leipzig Glossing Rules*. All glosses and other abbreviations created by us for this study specifically are given in the list of abbreviations. Except otherwise stated, all glosses and English translations are ours. Original glosses which do not correspond to the *Leipzig Glossing Rules* have been modified. No attempt has been made to homogenise orthographical variation. German examples for which no source is provided are constructed on the basis of our native-speaker competence.

(b) Place name
    bas~bassit   **ti** Oahu   *ngem*   **ti**   Luzon.
    CMPR~small   X   Oahu   but   X   Luzon
    'Oahu is smaller than Luzon.'
(c) Personal name
    na-gú~guápo   **ni** Robert Redford   *ngem*   **ni** Paul Newman.
    ADJR-CMPR~handsome   Y   Robert Redford   but   Y   Paul Newman
    'Robert Redford is handsomer than Paul Newman.'

There is no equivalent for X and Y in the English translation. In point of fact, both *ti* and *ni* belong to the article system of Iloko outlined in Table 1 based on Rubino (2011: 333) whose terminology is kept for convenience. What is recognisable immediately is the bipartition of the system into two major categories, namely that of non-personal as opposed to that of personal (cf. also Aikhenvald 2015: 89)

Table 1: Iloko articles (Rubino 2011: 333).

	non-personal			personal	
	singular		plural	singular	plural
	neutral	definite			
core	ti	diay	dagiti	ni	da
oblique	iti		kadagiti	kenni	kadá

According to Rubino (2011: 333), except in the vocative (cf. Section 2.4.2), personal nouns (= personal names) must always co-occur with the appropriate form of the personal articles whereas non-personal nouns drop the article when used predicatively. According to Espiritu (1984: 175–176), the non-personal core article can be dropped in existential constructions to indicate indefiniteness as in (13) below. Personal nouns cannot be made indefinite.

Thus, we state that there already are several instances of disagreement between non-personal and personal nouns. First of all, there are distinct forms of the article for the members of the two categories. The second difference is distributional in the sense that the non-personal article is blocked under predication whereas the use of the personal article is mandatory outside the vocative.[2] The distinction between neutral and definite article in the core singular is unique to non-personal nouns. Nothing similar can be found in the cells of the paradigm of personal nouns. There are morphological and morphosyntactic differences. Personal names behave differ-

---

[2] Example (7b) features a personal noun accompanied by the article *ni* in predicative function.

ently from other nouns. It is not possible to capture the behaviour of the former by way of using the rules which determine the behaviour of the latter and vice versa.

On the basis of the information given in Table 1, we can spell out the variables used in (1) as
- X = non-personal article neutral core singular = (glossed as) NPERS.SG and
- Y = personal article core singular = (glossed as) PERS.SG.

This, however, is not the end of the story Iloko can tell us. As transpires from the following exposition, the article system as reflected in Table 1 is incomplete. To understand why, we have to look at the oblique forms used for non-personal nouns. Espiritu (1984: 64) states that

> [r]eal locations, like places, are usually introduced by *ditoy, dita,* and *idiay*. Locations which are more abstract such as family and mathematics are usually introduced by *iti*. This determiner has many uses in Ilokano, all of which can be broadly characterized as locative, or having to do with location.

The quote reveals that Table 1 fails to acknowledge the existence of a distinct oblique article employed for expressions which refer to what Espiritu calls locations. What is meant by this term becomes slightly more transparent in the next quote from Espiritu (1984: 144) who claims that

> [a] location need not be a place name, such as *Baguio*, or a common noun such as *balay* (house), a location is used as the locative determiner before a personal noun.

Baguio is the name of a city situated on Luzon. It is a genuine place name. In contrast, *balay* 'house' does not refer to a geo-object with fixed coordinates on the world map. It fails to fulfil the criterion of mono-reference (Nübling et al. 2015: 17–19) and thus seems to be a bona fide common noun. Superficially, what Espiritu has to say about personal nouns in the above quote sounds confusing. However, the confusion is easy to clarify since the locative determiner is identical to the oblique personal article. The way Espiritu describes the choice of locative determiner with non-personal nouns is suggestive of a division in two on the basis of the opposition [concrete] ≠ [abstract]. The examples in (2) do not corroborate this hypothesis. Articles which are not important for the ensuing discussion are underlined. The important ones appear in boldface.

(2) Iloko [Austronesian, Northern Luzon]
    (a) place name (Espiritu 1984: 44)
        Nai-yanak   <u>da</u>   Pablo   ken   Luna   **idiay**   Vigan.
        PFV-born   <u>PERS.PL</u>   Pablo   and   Luna   **X**   Vigan
        'Pablo and Luna were born **in** Vigan.' [O.T.]

(b) common noun I (Espiritu 1984: 138)
Ania <u>*ti*</u>   *in-arámid-yo* **idiay** *eskuéla ita?*
what NPERS.SG PFV-work-2PL X school now
'What are you doing **at** school at the moment?'
(c) common noun II (Espiritu 1984: 152)
in-ála-da   <u>*dagiti*</u> *aláhas ken kuarta nga adda*
PFV-take-3PL.ERG NPERS.PL jewelry and money LK EXIST
iti   *aparador.*
NPERS.OBL dresser
'They got <u>the</u> jewelry and money that were **in the** closet.' [O.T.]
(d) personal name (Espiritu 1984: 145)
ag-telépono-ka   *kada*   *Julian*
INTR-telephone-2SG.ABS PERS.PL.OBL Julian
'Call up Julian's **place**!' [O.T.]

(1d) is in line with what the literature assumes as to the behaviour of personal names functioning as Ground in a spatial situation,[3] i.e., the expected oblique personal article is made use of. Similarly, (1a) and (1b) meet the expectations because a place name – *Vigan* (capital city of Ilocos Sur on Luzon) – and a concrete common noun – *eskuéla* 'school' – require the presence of *idiay*. *Idiay* is glossed as X in these examples because the form is absent from Table 1. Since *idiay* fills the same slot as *kada* in (1d) we assume that the two syntactic words share some functions. In this case, both are employed to mark the oblique. Given that the functional equivalence holds, *idiay* must be given the same status as any other oblique article in the system.

The above hypothesis according to which place names and concrete common nouns behave identically is at odds with (1c) though. The Ground in the spatial situation is the concrete common noun *aparador* 'dresser'. Contrary to expectation, there is no trace of *idiay* in the sentence. What we find in its stead is *iti*, supposedly the oblique article of abstract common nouns. Accordingly, the originally assumed opposition [concrete] ≠ [abstract] becomes doubtful. In point of fact, it can be shown that there is no such opposition that regulates the choice of the oblique article of non-personal nouns in the first place. To prove our point, we have collected all instances of *idiay* and *iti* which show up on the 290 pages of Espiritu (1984). The token frequencies do not matter. What counts for what we are going to argue is the

---

[3] For the purpose of this study, we make use of the established terminology of studies on motion events (Talmy 1985). Since the basic notions are especially important with regards to the case studies presented in Section 5, the necessary definitions of the concepts will be provided in the introduction to Section 5 below.

combinability of *idiay* and *iti* with different kinds of non-personal nouns. In spatial situations, both *idiay* and *iti* neutralise the distinction of Place ≠ Goal.

For a start, we sketch the co-occurrences of *iti* with NPs over which it has scope as a determiner and oblique marker. *Iti* is frequently attested in contexts which cannot be characterised as being genuinely spatial. From contexts like those in (3) and further non-spatial ones, *idiay* is strictly excluded.

(3) Iloko [Austronesian, Northern Luzon]
    (a) language (Espiritu 1984: 293)
        *ania   ti   átis **iti**   Inggles Nána?*
        what NPERS.SG átis **NPERS.OBL** English granny
        'What is *átis* **in** English, granny?' [O.T.]⁴
    (b) abstract concept (Espiritu 1984: 58)
        *sinno ti   maestro-m   **iti**   Matemátiks?*
        who   NPERS.SG teacher-POSS.2SG **NPERS.OBL** mathematics
        'Who is your teacher **in** maths?'
    (c) event (Espiritu 1984: 200)
        *k<um>úyog-ak-to   láengen **iti**   tour*
        <INCH>go_together-1SG-FUT just   **NPERS.OBL** tour
        'I will just participate **in** a tour.'
    (d) time (Espiritu 1984: 83)
        *mapan ag-bása   ni   Luz **iti**   rabii*
        go   INTR-study PERS.SG Luz **NPERS.OBL** evening
        'Luz goes to school **at** night.' [O.T.]
    (e) social network (Espiritu 1984: 52)
        *sinno met ti   Amerikáno **iti**   pamiliáyo?*
        who   also NPERS.SG American **NPERS.OBL** family
        'And who's **the** American **in** your family?'
    (f) human Ground noun (Espiritu 1984: 148)
        *ited-ko   daytoy   sábong **iti**   ubing*
        give-1SG.ERG DEM.PRX flower **NPERS.OBL** child
        'I will give this flower **to the** child.'

Like in (2c), there are many spatial situations in which *iti* serves to introduce the noun (or NP) which functions as Ground, a typical example being (4).

---

4 *Átis* is a kind of tropical fruit (sweetsop) (Rubino 2000: 68).

(4) Iloko [Austronesian, Northern Luzon] (Espiritu 1984: 228)
    nanganda   **iti**   restawran   **ti**   Insik
    PFV:eat   **NPERS.OBL**   restaurant **NPERS.SG** Chinese
    'S/he has eaten **in** the Chinese restaurant.'

Further nouns combining with *iti* in spatial situations featured in our source are listed alphabetically in (5). The bracketed numbers refer to the pages in Espiritu (1984). Non-spatial usages of *iti* like combinations with *koriénte* 'electricity' (181), *radio* 'radio' (188), *sueldom* 'your salary' (207), and *tíbi* 'TV' (130, 159) are also frequently attested.

(5) Nouns combining with *iti* in Espiritu (1984)
    aparador 'dresser' (152), *bolsána* 'his/her pocket' (227), *bus* 'bus' (150), *freeway* 'freeway' (181), *karro(da)* '(their) car' (150, 152, 153), *lamisáan* 'table' (147), *libro* 'book' (174), *manga* 'mango tree' (109), *pagparadáan* 'parking lot' (181), *sine* 'cinema' (92)

In addition to these instances of *iti* + NP, *iti* is frequently involved in the formation of complex prepositions of the type
– *iti* + RELATIONAL NOUN [+ *ti* + NON-PERSONAL NOUN]$_{\text{COMPLEMENT}}$

such as those given in (6).

(6) Complex prepositions (Espiritu 1984: 174)
    iti rabaw ti lamasáan   'on the table'   (*rabáw* 'upper side')
    iti sírok ti katre   'under the bed'   (*sírok* 'space under something')
    iti ábay ti tugaw   'beside the chair'   (*ábay* 'side')

Idiay does not partake in the formation of complex prepositions.[5] Its uncontested domain is genuine place names which can be names of countries as in (7a), names of cities as in (7b), names of quarters as in (7c), or names of institutions as in (7d).

(7) Iloko [Austronesian, Northern Luzon]
    (a) country (Espiritu 1984: 144)
        i-pawit-ko   dagitoy   bangbanglo   **idiay**   Pilipínas
        VR-something_sent-1SG.ERG   DEM.PRX.PL   perfume   **X**   Philippines
        'I will send these perfumes **to the** Philippines.' [O.T.]

---

5 In Espiritu (1984: 144), the adverbial *idiay uneg* 'on the inside' is attested whereas Rubino's (2000: 398) dictionary gives *iti uneg* as complex preposition meaning 'inside'.

(b) city (Espiritu 1984: 216)
   ni   Lawrence Olivier   ti   ka-laing-an   nga   artista
   PERS.SG   Lawrence Olivier   NPERS.SG   SUP-aptitude-SUP LK   artist
   idiay   Hollywood
   X   Hollywood
   'The best artist **in** Hollywood is Lawrence Olivier.'

(c) quarter (Espiritu 1984: 73)
   pag-trab~trabahua-m–   *idiay*   downtown
   LOC-CONT~work-POSS.2SG   X   downtown
   'Where are you working? – Downtown.'

(d) institution (Espiritu 1984: 44)
   Agka-eskuelá-an-da   *idiay*   McKinley High School
   REC-school-REC-3PL   X   McKinley High School
   'They are classmates **at** McKinley High School.' [O.T.]

Never does *iti* replace *idiay* as oblique article of place names. This rule also holds for the cases presented in (8).

(8) Place names combining with *idiay* in Espiritu (1984)
Chicago (216), *Dole Pineapple Company* (44), *Kalihi* (130),[6] *Kapitolio* 'Capitol' (229), *Sampagita Café* (214, 215), *Tsína* 'China' (137), *Vigan* (44), *Waialua* (44),[7] *Waialua Piggery* (219)

Idiay has the monopoly not only for combinations with genuine place names but also for those with a number of common nouns such as rooms in a building as in (9a) and anonymous institutions as in (9b).

(9) Iloko [Austronesian, Northern Luzon]
   (a) room (Espiritu 1984: 231)
       adda   kama   *idiay*   sála?
       EXIST   bed   X   parlour
       'Is there a cushioned bed **in the** parlour?'
   (b) institution (Espiritu 1984: 228)
       napan   ni   Isabel   *idiay*   pag-siapiíng-an
       PFV:go   PERS.SG   Isabel   X   LOC-shopping-LOC
       'Isabel went **to the** shopping mall.'

---

6 *Kalihi(-Palama)* is a quarter of Honolulu.
7 *Waialua* is a village on O'ahu (Hawaii).

In (10), further common nouns which combine with *idiay* are identified.

(10) Common nouns combining with *idiay* in Espiritu (1984)
      bangko 'bank' (150), *kábinet* 'cabinet' (165), *kapiteria* 'cafeteria' (138, 150), *kusína* 'kitchen' (130, 132), *laybrari* 'library' (138, 147), *tiendáan* 'store' (150)

It is cross-linguistically rather common that the boundaries between names and nouns are fuzzy in the sense that certain common nouns structurally behave like place names whereas others behave like personal names. The latter is the case, for instance, with kinship terms in Iloko (and many other languages across the globe). According to Espiritu (1984: 6)

> [w]ords like *Nánangko* (my mother), which name kinship relations, are personal nouns in Ilokano, as though they were people's names.

The reason for the inclusion of these kinship terms in the class of personal nouns is functional because both kinship terms and names can be used to address interlocutors. Analogously, common nouns like those presented in (10) are functionally similar to genuine place names because they serve as labels for prototypical places in the surroundings and experience of people. The parts of buildings referred to, for instance, are those which belong to the interlocutors' common ground – the cafeteria and the library in the school they attend daily, the kitchen at their home, the bank and the store in the vicinity, etc. The common nouns used to refer to these locations are like place names because, unless otherwise specified, they are mono-referential. Mono-referentiality can be achieved by way of adding a possessor[8] to the common noun as in (11) which involves *lugar* 'place'.

(11) Iloko [Austronesian, Northern Luzon] (Espiritu 1984: 181)
      *Naka-tag~tagari   ngamin   **idiay**   lugar-mi*
      INTS-INTS~noisy   because   X   place-1PE
      'Because it is very noisy **at** our place.'

In the vast majority of cases, these common nouns (as "honorary" members of the class of place names) designate places where human beings can be or move to. This also includes metonymic extensions of the spatial concept of place to events which

---

8 The concepts, notions, and terms which are crucial for the interpretation of *Special Onymic Grammar* in the functional domain of possession in Section 4 correspond to those introduced or propagated by Heine (1997: 33–41 and 143–186). They are explained in more detail in Section 4.5 below.

take place at a given location such as *padayá* 'feast' (148, 217, 218, 219) as shown in (12).

(12) Iloko [Austronesian, Northern Luzon] (Espiritu 1984: 218)
 ania ti ado **idiay** padaya?
 what NPERS.SG many X feast
 'How many are **at the** party?'

A notable exception to the above is *kábinet* 'cabinet' (165) whose combinability with *idiay* is in conflict with that of *aparador* 'dresser' (152) since both refer to pieces of furniture which are employed as containers for concrete objects – and not for humans. This is slightly different with *tugaw* 'chair' (174, 175) (cf. (13)) because people usually sit on chairs whereas they normally do not pass their time inside dressers or cabinets.

(13) Iloko [Austronesian, Northern Luzon] (Espiritu 1984: 175)
 adda libro **idiay** tugaw
 EXIST book X chair
 'There is a book **on** the chair.'

Admittedly, the explanation for the use of *idiay* with *tugaw* 'chair' is ad hoc. To render it more robust more data are needed. The same holds for those cases where *idiay* and *iti* compete with each other for combining with certain common nouns. There are altogether thirteen instances of the common noun *balay* 'house' combining with an oblique article. Twelve times the article comes in the shape of *idiay* as in (14a) but only once does *iti* occur as in (14b).

(14) Iloko [Austronesian, Northern Luzon]
 (a) *idiay* (Espiritu 1984: 82)
  kanáyon a map~mapan ni Josie **idiay** balay-da
  always LK FREQ~go PERS.SG Josie X house-POSS.3PL
  'Josie is always going **to** their house.' [O.T.]
 (b) *iti* (Espiritu 1984: 131)[9]
  S<imm>agpát **iti** balay-da
  <PFV>go NPERS.OBL house-POSS.3PL
  'She went up **to** their house.'

---

[9] Finite verb added by us.

There is no apparent explanation for this variation since in both cases *balay* hosts the possessor suffix of the 3rd person plural and serves as the Ground of a spatial situation in which motion towards a Goal is described. In any case, the co-occurrence of *iti* and *balay* is exceptional. What about the competition of *idiay* and *iti* in combination with *siled* 'room'? Half a dozen tokens of this common noun are attested in the oblique in our source. Four out of six times, *idiay* is used as in (15a) whereas *iti* is attested twice as in (15b).

(15) Iloko [Austronesian, Northern Luzon]
    (a) *idiay* (Espiritu 1984: 175)
        adda **idiay** *siled* <u>*ti*</u>   *púsa*
        EXIST **X**   room NPERS.SG cat
        '<u>The</u> cat is **in the** room.'
    (b) *iti* (Espiritu 1984: 229)
        awan   <u>*ni*</u>   *Merle* **iti**   *nalabbága* *a* *siled*
        NEG.EXIST <u>PERS.SG</u> Merle **NPERS.SG.OBL** ADJR:redness LK room
        'Merle is not **in the** red room.'

In all cases where *idiay* accompanies *siled*, the common noun takes no further attributes whereas in both cases where *iti* is used, *siled* is either modified by an adjective or quantified by a cardinal numeral (= *iti maysa a siled* 'in one of the rooms' (152)). Whether the structural complexity of the NP plays a role in the choice of the oblique article is a question that cannot be answered on the limited basis of data we have access to. The next case of competition between the two oblique articles poses similar problems. In (16), we observe variation in the use of *idiay* and *iti* in combination with the synonymous common nouns *eskuela* 'school' (138, 143, 193) and *pageskueláan* 'school' (181).

(16) Iloko [Austronesian, Northern Luzon]
    (a) *idiay* (Espiritu 1984: 193)
        p<in>erdi   <u>*dagiti*</u>   *na-lóko*   *nga* *estuediante*
        <PFV>destroy NPERS.PL ADJR-fool LK   student
        <u>*dagiti*</u>   *relo*   **idiay** *eskuéla*
        NPERS.PL clock **X**   school
        '<u>The</u> troublesome students destroyed <u>the</u> clocks **in the** school.' [O.T.]
    (b) *iti* (Espiritu 1984: 181)
        asideg *ngamin* **iti**   *pag-eskuelá-an*
        near   because **NPERS.OBL** LOC-school-LOC
        'Because it is near **the** school.'

Once again, we lack sufficient data to draw conclusions. Cross-linguistically, it frequently makes a difference whether a given common noun is ambiguous in the sense that it may either refer to a concrete building or to the institution (often hosted in the building). This disambiguation, however, does not apply to the cases discussed above. The last pair of sentences in (17) involves the common noun *tawa* 'window' which is attested once with *idiay* and twice with *iti* in the source text.

(17) Iloko [Austronesian, Northern Luzon]
    (a) *idiay* (Espiritu 1984: 179)
        ag-tarayka-yo **idiay** *táwa*
        INTR-elope-2PL **X** window
        'Get out **through** the window!'
    (b) *iti* (Espiritu 1984: 152)
        Kalpasanna imm-úli-da **iti** *táwa*
        afterwards PFV-climb-3PL **NPERS.OBL** window
        'Afterwards, they climbed **through** the window.' [O.T.]

The impossibility of deciding these problematic issues notwithstanding, one thing seems to be clear, namely that genuine place names never take the oblique article *iti* and the bulk of the common nouns does not combine with *idiay*. For a probably limited number of common nouns, both options appear to be licit with some of these nouns preferring *idiay* over *iti* and others displaying the opposite preference. Genuine place names constitute the core of a potentially extendable third kind of nouns distinct from non-personal common nouns on the one hand and personal nouns on the other hand. Their main distinctive feature is the obligatory use of an oblique article which is not identical to that employed for non-personal common nouns. On this basis, we can now define the mysterious X which hitherto served as gloss for *idiay* as

– oblique singular place name = (glossed as) PLC.OBL.[10]

The existence of a plural counterpart is not assumed although personal nouns are equipped with distinct plural forms of their articles.

What makes this additional oblique article special is the fact that speakers of Iloko may take their pick from a set of three because apart from *idiay* there are

---

10 Only in passing do we refer to the possibility of using this analysis of Iloko *idiay* to explain the still largely obscure diachrony of the Chamorro place-name article *iya* /idza/ which we assume to be a cognate of the Iloko equivalent (Stolz 2020). The voiced affricate /dz/ can be traced back to a sound change which transformed the sequence /d/ + /i/ to /dz/. At the same time, the Chamorro parallel supports the idea that Iloko *idiay* is functionally the same, i.e., a place-name article.

also *ditoy* and *dita* which together form the ternary paradigm of deictics of Iloko termed locatives by Rubino (2000: lxxvi). Proximal *ditoy* 'here', medial *ditá* 'there', and distal *idiáy* (also: *sadiáy*) 'there (out of sight)' "are also used to introduce locative nouns with reference to the location of the speaker" (Rubino 2000: lxxvi). It is therefore possible to locate people in three different ways in Ilocos[11] in correlation to the whereabouts of the speaker (and his/her interlocutor) as shown in (18) based on Espiritu (1984: 222).

(18) Iloko [Austronesian, Northern Luzon]

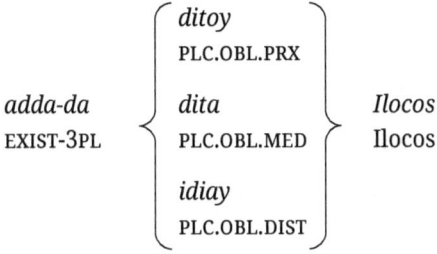

'They are here / there / over there in Ilocos.'

If we omit the place name from example (18), we get perfectly well-formed sentences as shown in (19).

(19) Iloko [Austronesian, Northern Luzon]

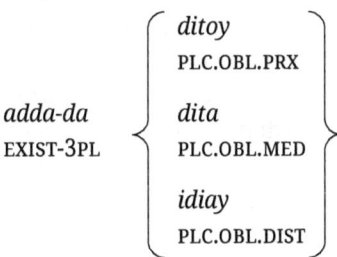

'They are here / there / over there.'

On account of this possibility, one might want to object to the analysis of the deictics as oblique place-name articles. However, nothing speaks against classifying them as polyfunctional grammatical elements. Moreover, they are employed exactly

---

11 *Ilocos* (aka *Region I*) is the name of an administrative district on the north-western coastline of Luzon.

in those contexts in which personal nouns require the presence of their oblique article *kenni* and non-personal common nouns are preceded by *iti*. The above deictics occupy exactly the same slot as their personal and non-personal equivalents. Like in the case of the personal articles, the place-name article cannot be dropped either. Thus, there is good reason to reorganise Rubino's article system as of Table 1 to yield a very different shape in Table 2.[12]

Table 2: Iloko articles (revised).

	common			place			personal	
	singular		plural				singular	Plural
	neutral	definite						
core	*ti*	*diay*	*dagiti*	*ti*			*ni*	*da*
oblique	*iti*		*kadagiti*	*ditoy*	*dita*	*idiay*	*kenni*	*kadá*
				PRX	MED	DIST		

As to the formal side of the expressions, the paradigm is relatively homogeneous. If we add the appropriate nouns to the articles in their respective cells, we always get binary combinations whose left slot is occupied by the article and the right slot by the noun. Most of the word forms in this paradigm can be morphologically decomposed. The common-noun article *ti* is recognisable also in the oblique *i-ti* just like the personal article *ni* resurfaces in the oblique *ken-ni* in the singular and *da* in *ka-da* in the plural, etc. What is especially noteworthy is the relation of the distal oblique form of the place-name article *idiay* to the definite singular common article *diay*. The former contains the same initial *i-* as found in *iti*, i.e., this *i-* is – perhaps only diachronically – the oblique case marker. Since there is no overt case-marking for the core case, it is legitimate to ask what function the remaining phonological string might have. In the case of common nouns, definiteness marking certainly comes into play because bare nouns are indefinite. Neither place names nor personal names can be stripped of their articles (except in the vocative). However, names are commonly considered to be inherently definite. With reference to names in general, Van Langendonck and Van de Velde (2016: 22) state that

---

[12] For lack of sufficient data, we refrain from assuming a further class of nouns, namely those which optionally behave like place names or common nouns. The integration of this class of vacillating nouns would of course render the paradigmatic survey even more complicated than Table 2.

> [s]ince names are inherently definite, the addition of an overt definiteness marker is superfluous, and definite markers are often used to express notions other than definiteness.

The scenario described in this quote does not, however, apply to Iloko. There are two factors that conspire. On the one hand, the articles are responsible for distinguishing core from oblique. This motivates the binary vertical paradigm. On the horizontal plane, number distinctions are crucial for common nouns and personal nouns but not for place names. We assume that in lieu of superfluously coding definiteness, the articles of place names and personal names are principally in charge of classification, i.e., they inscribe the nouns they accompany into a subclass of nouns. They do not conform to the criteria for the definition of articles put forward by Becker (2021: 32–56). With reference to data from another Austronesian language, Ughele, which uses *e* with personal names and *na* with common nouns, Becker (2021: 41) excludes markers of this kind from her typology of articles because

> their function is not linked to referentiality, but [. . .] they mark other lexical properties of the referent (human vs non-human) besides their syntactic function as a determiner.

We doubt that the assumed opposition of human vs non-human adequately captures the role of these markers.[13] Still, we agree with Aikhenvald (2015: 122) that the term article is most probably a misnomer for the word-class markers of place names and personal names. Yet, we will keep the incriminated term most of the time because it is firmly established. Always replacing it for good with labels like place-name marker and personal-name marker (which would be two sub-classes of onymic markers) is no solution either because, to the best of our knowledge, in the literature on classifiers and genders, markers of this kind are not mentioned. Yet, these alternative terms occasionally pop up in the subsequent paragraphs without invoking a meaning difference in contrast to place-name article.

As is evident from Table 2, the paradigms of personal names, place names, and common nouns are not identical. There is a high degree of heterogeneity. This heterogeneity manifests itself in a number of mismatches in the sense of Canonical Morphology (Corbett 2007: 30) whose principles are discussed separately in Section 2. For the time being, it suffices to provide a list of those phenomena which

---

13 Becker (2021: 40–41) takes Ughele only as illustration of a constellation of facts which is typical for many Austronesian languages. In these languages, it is usually the case that the common article is used in combination with non-proprial human nouns whereas the proprial article is restricted to names. The distinction is thus between proper names (with human referents as default) and common nouns, be they human or other.

show that the integration of names into the paradigm of nouns yields a colourful picture:
- **inflectional classes**: common nouns, personal nouns, and place names constitute distinct inflectional classes in the sense that the cells of the paradigms are filled with different grammatical markers for identical functions (such as *i-* for the oblique singular of common nouns but *ken-* for the oblique singular of personal nouns);
- **overdifferentiation**: only common nouns distinguish neutral from definite forms in the singular of the core case (*ti* vs *diay*) whereas only place names differentiate between three deictic forms in the oblique (*ditoy* ≠ *dita* ≠ *idiay*);
- **syncretism**: common nouns and place names display the same marker of the core case *ti* (= neutral for common nouns);

Furthermore, place names use markers which belong to a different word-class from nouns (deictics) to form the oblique. In contrast to common nouns and personal nouns,[14] place names give no evidence of plural number. For both types of names, the use of the onymic marker is obligatory whereas the common-noun article alternates with zero to distinguish definite from indefinite. Personal nouns and place names are deeply involved in bringing about structural mismatches which crucially shape the paradigms of "nouny" word-classes of Iloko. Table 3 surveys the categories which are relevant for common nouns, personal nouns, and place names in Iloko.

Table 3: Categories relevant for nouny word-classes in Iloko.

	case	class	number	definiteness	deixis
common	yes	(yes)	yes	yes	no
person	yes	yes	yes	no	no
place	yes	yes	no	no	yes

Grey shading identifies those cases where names diverge from common nouns. Place names are involved in divergence three times. In two of these cases, there is also a behavioural split between the two kinds of names. This means that personal nouns and place names cannot simply be lumped together. Personal nouns share more features with common nouns than they have in common with place names.

---

14 Sentences (2a) and (2d) contain personal names in combination with plural markers. Plural marking occurs under coordination of several personal nouns and to form a kind of associative plural meaning X + family – a category which will be of interest again in Section 2.3.1.

However, Iloko is not alone. With reference to the Austronesian languages of Asia and Madagascar, Himmelmann (2005: 145) states that "[t]he personal name markers are always obligatory" whereas articles of common nouns may be omitted in several languages. In the western section of the Austronesian speech territory, Tagalog, Cebuano, Kimaragang, Malagasy, Buol, Mori Bawah, Kambera, to name only a few, boast personal-name markers alongside common articles. For the Oceanic languages in the eastern part of Austronesia, Lynch et al. (2011: 38) report that

> [m]any Oceanic languages have articles that precede a noun phrase. These often make a distinction between singular and plural, and between common and proper, and sometimes make a more fine-grained set of semantic distinctions than that.

Evidence of personal-name markers alongside common-name articles can be found among many other Oceanic languages in Bali-Vitu, Kaulong, Siar, Taiof, Roviana, Gela, Arosi, Tamambo, Cèmuhî, Rotuman, Nadrogā, Marquesan, etc. In all of these languages, personal names differ from common nouns. These open lists strongly suggest that we are not dealing with a marginality – SOG is attested far too often across Austronesian to be ignored. In the empirical part of this study, we will come back to some of the above members of the Austronesian language family without, however, pretending that SOG is the privilege of Austronesian. Sections 3–5 are meant to prove that SOG is cross-linguistically widespread and by no means restricted to the presence of dedicated onymic markers. We summarised this prelude for Iloko with the following formula which reflects the above paradigmatic differences:
- COMMON NOUNS ≠ PLACE NAMES ≠ PERSONAL NAMES.

This simple formula serves as a guiding line for the project we are conducting.

1.2 The study in a nutshell

The foregoing sketch of SOG in Iloko circumscribes the kind of situations we are going to hunt in the envisaged project. We collect data from as many languages as possible which prove that names do not behave like the average common noun. In the next step, the structural facts are systematised in order to facilitate the discovery of similarities across the languages of the sample. These similarities are then evaluated from the point of view of functional typology. To this end, several decisions have to be taken which guarantee the feasibility of our endeavour.

First of all, we limit the investigation to personal names and place names because of two reasons. On the one hand, if there is any information on names in descriptive grammars at all, this information covers personal names and sometimes, but less often also place names whereas the bulk of the numerous other name classes is tacitly passed over for the most.[15] On the other hand, personal names and place names rank high on the prototypicality scale proposed for name classes by Nübling et al. (2015: 104). For practicality, we turn a blind eye to further subdivisions of the name classes under inspection, meaning: we do not further differentiate between given names, family names, nicknames, etc. as all of them are personal names for the purpose of our project;[16] similarly, settlement names, country names, island names, etc. are indiscriminately registered as place names. The absence of a section explicitly dedicated to names, e.g. in grammars written in the LDS format,[17] at least partly explains why reliable information on SOG is very hard to come by for a sizable sample of languages.[18] As comes to the fore in Sections 3–5, by far not all members of a given name class behave the same in terms of morphology and morphosyntax. This means that SOG features might surface only in a small subset of the name classes under scrutiny.

Second, the terminology varies considerably in the extant literature. In Section 1.1, we have deliberately employed the terms which are on offer in the descriptions of Iloko. What the authors call (personal) nouns there is labelled (personal) name or anthroponym elsewhere. Similarly, place names may also go by the alias toponyms, etc. To avoid confusion, we homogenise the terminological diversity by way of introducing the following labels and abbreviations thereof. The term *anthroponym* (shorthand: ANTH) replaces personal name/noun. If it is necessary to refer separately to the class of common nouns which (optionally or obligatorily) behave like ANTHs, these common nouns are referred to as *anthronouns* (= ANTH-N). The artificial compound *anthronoun* is patterned on the model of Haspelmath's (2019: 330) felicitous proposal *toponoun* (= TOPO-N) for those common nouns which morphosyntactically behave like place names. The latter are named *toponyms* (= TOPO)

---

15 Van Langendonck (2007: 186–255) lists at least sixteen different name classes. Nübling et al. (2015: 98–336) discuss six major name classes with forty-six subdivisions. The *Oxford Handbook of Names and Naming* (Hough 2016) contains sixteen chapters each dedicated to a distinct name class, etc.
16 The referent of a personal name need not be human since personification makes it possible to transfer personal names to non-human referents.
17 The LDS Section 16.1 is normally reserved for the operational definition of nouns. Some grammarians mention proper nouns and compare them to common nouns whereas in many cases no light is shed upon names and their structural behaviour even if the language described in the grammar attests to SOG.
18 Symptomatically, in an otherwise excellent grammar of the Bodish language Kurtöp, the index tells the reader under the entry proper name that this category is simply "not discussed" in the text (Hyslop 2017: 455).

in this study. Common nouns are not affected by our terminological reorganisation although they are normally referred to with the abbreviation COMM. ANTH and TOPO together form the class of *proper names* (= PROP) to which PROP-N (short for *proper noun*) – constituted by ANTH-N and TOPO-N – can be added.[19] The formula introduced in Section 1.2 henceforth appears in this shape:
– COMM ≠ TOPO ≠ ANTH.

Third, Anderson (2007: 213–261) and Van Langendonck (2007: 125–168) focus on the syntactic properties of names. Nübling et al. (2015: 64–92) additionally take account of phenomena in segmental and suprasegmental phonology, derivational and inflectional morphology, morphosyntax, and graphematics. These previous ground-breaking works notwithstanding, the domain of SOG is still largely understudied especially if it comes to comparing cross-linguistic evidence on a grand scale. It is advisable therefore to restrict the scope of the inquiry to one or two of the levels of grammar in lieu of trying to address all interesting facts in one go. This is why we concentrate our efforts on inflectional morphology and selected morphosyntactic aspects of SOG. Special attention is paid to those phenomena which block the application of COMM-rules to PROPS. The necessity of formulating extra-rules for PROPS is indicative of the existence of structural mismatches. Only in passing do we mention additional phenomena from other domains (in Section 3).

Fourth, we opt for working on the basis of a convenience sample in order to provide illustrative examples of the manifestations of SOG in a wide variety of languages.[20] We are primarily interested in qualitatively describing the principles of SOG. It is therefore of minor importance for us to determine quantitatively how many languages world-wide attest to SOG. There is thus no need for a balanced sample which also comprises languages without SOG features. On the assumption that PROPS as a category are panchronically universal, any language at any time and at any place might potentially display signs of SOG, develop or lose them. Having said this, we emphasise that we do not claim there is SOG no matter where and when you look. To avoid speculating unreasonably, we exclusively present and discuss water-tight evidence of SOG. As will become evident in the empirical Sections 3–5, finding the pieces of evidence frequently requires using the linguistic looking-glass or even the linguistic microscope since dedicated markers are often wanting.

---

19 The abbreviations introduced in this paragraph are used only in the main body of the text and footnotes. They are avoided in section headings and direct quotes.
20 Owing to the nature of the sources we use for this project (cf. below), the term language should be understood as referring to doculects in the sense of Cysouw and Good (2013). Furthermore, language serves as cover term of standard and nonstandard varieties alike.

Fifth, given the lamentable scarcity of available cross-linguistic data, the sources from which the data are drawn for this study represent a variety of genres. The usual descriptive-linguistic material such as grammars, dictionaries, specialised studies, etc. are complemented by pedagogical grammars, text anthologies, electronic corpora, etc. and occasionally by personal communication with language experts. Only in Sections 4–5 do we present data taken from primary sources. The project is still in its initial phase. Therefore, it makes sense first to amass (putative) evidence of SOG and to subsequently clear up the data-base by way of deleting doubtful cases and evident errors. Since we assume that SOG is not restricted to a given epoch, we accept data from different historical stages of the sample languages. In this sense, we approach SOG not only synchronically but also diachronically (although it might not always be possible to trace the succession of developments in time which have brought about a given situation in a given language).

Sixth, the project is situated in the larger context of functional typology whose usual comparative methodology is applied. The approach adopted here is framework-free in the sense of Haspelmath (2010). Nevertheless, we largely orient ourselves along the lines of ideas and postulates of Construction Grammar (Booij 2013). For the analysis and evaluation of the morphological and morphosyntactical structures, the principles and notions of Canonical Typology are made use of (Brown and Chumakina 2013, Corbett 2013). For everything that is directly associated with the concept of SOG, Nübling et al. (2015) are our main frame of reference (with the occasional sideways glance at Anderson (2007) and Van Langendonck (2007)).

The study is organised as follows. In Section 2, a short review of the extant literature in the domain of the grammar of names is given with a view of determining the proper place of SOG in the ongoing debate. It is in Section 2 that we disclose the working hypothesis for this study. Section 3 contains an annotated assortment of statements in descriptive-linguistic texts on structural facts which might turn out to be of relevance to the study of SOG but are not sufficiently elaborated upon to allow for the formulation of robust generalisations. In contrast to this (incomplete) catalogue of largely isolated data, Sections 4–5 are each divided into several subsections which host more detailed case studies.[21] Those which revolve around ANTH(-N) alone can be found in Section 4 whereas case studies which focus

---

[21] For the empirical case-studies, possession and spatial relations are chosen on purpose because both of these domains have been studied in-depth for many years by members of the Bremen research team albeit for the most without special connection to PROPS. It suffices to pick out two representative major publications to prove that research on possession (Schuster 2020) and spatial relations (Nintemann et al. 2020) is firmly established at the Institute of Linguistics at the University of Bremen/Germany.

on TOPO(-N) are presented in Section 5. The cross-linguistically recurrent patterns are evaluated with reference to functional criteria in Section 6. The study closes with Section 7 in which the tenability of the working hypothesis is determined. In the same final section, we put forward the conclusions and outline possible topics for follow-up studies.

2 The Grammar of Names vs Special Onymic Grammar

In the subsections of this section, selected milestones of the research on PROPS are reviewed summarily. The scholars whose contributions to the topic of interest are featured below have authored many more studies which, in a different context, would be worth discussing in detail. We will come back to some of their ideas (skipped in this section) at the appropriate places in the subsequent sections. For accounts of the history of thought and recent developments in the domain of the linguistics of PROPS, the reader is referred to Anderson (2007: 71–210), Van Langendonck (2007: 24–64), Schlücker and Ackermann (2017), Schlücker et al. (2017), Handschuh and Dammel (2019), Kempf et al. (2020), Levkovych and Nintemann (2020), and Caro Reina and Helmbrecht (2022).

To properly assess the morphological and morphosyntactic phenomena which distinguish PROPS from COMMS and ANTHS from TOPOS, we take as our guideline the catalogue of morphological mismatches featured in Corbett (2007: 30) and elaborated upon in a plethora of publications authored by members of the *Surrey Morphology Group* since the early 1990s. Mismatches instantiate noncanonicity, i.e., a given structural pattern fails to meet the expectations formulated on the basis of a set of principles which require of a paradigm that ideally all cells are occupied by one word-form whose internal composition is the same for each inflected word reflecting diagrammatism and whose lexical component remains the same throughout the paradigm (Corbett 2007: 23–24). Following Corbett (2013), we apply the originally morphological criteria of canonicity to morphosyntactic constructions by way of treating sets of functionally interrelated multi-word constructions as paradigms along the above lines. The different kinds of mismatches attested in the languages we inspect will be dutifully identified one by one throughout the empirical parts of this study.

Before we can start to analyse concrete data, however, we need to review what our predecessors in the domain of the grammar of PROPS had to say with regard to the relationship between PROPS and COMMS.

2.1 What onomasticians are less interested in

The traditional home of PROP studies is onomastics whose venerable history, for obvious reasons, cannot be recapitulated here (Eichler 1995). The brief reference to two monumental handbooks of the onomastic science must suffice. We start with the chronologically earlier HSK volume *Namenforschung* (Eichler et al. 1995). In his short survey article, Bauer (1995: 15) claims that

> [d]a Eigennamen zunächst einmal Sprachzeichen sind, ist zweifellos die Sprachwissenschaft diejenige Disziplin, der die Untersuchung von Namen in erster Linie obliegt.[22]

Accordingly, the quoted author provides a long list of those branches of linguistics which, to his mind, are relevant for the study of PROPS. It is telling that cross-linguistic and typological approaches do not turn up on his list. In point of fact, most of the articles in Eichler et al (1995) focus on individual languages or on a small set of mostly (Indo-)European languages. It is therefore doubtful in many cases whether the conclusions drawn on this limited empirical basis can be generalised. The data discussed in our study reveal a much richer phenomenology of the grammar. We strongly emphasise that any kind of (certainly unintended) selective Eurocentrism has to be avoided if one aims at putting forward cross-linguistically valid hypotheses about the grammar of PROPS.

Furthermore, all levels of grammar are mentioned including syntax which, the author erroneously assumes, is treated in a separate dedicated article in the same handbook. However, the Section *Namengrammatik* of the handbook does not host a separate chapter on syntax. In Kolde's (1995: 407–408) survey of the grammar of PROPS, there is little more than a column in which syntactic issues are summarily addressed. The term morphosyntax is absent from this article and the entire section to which it belongs. The impression resulting from the little attention that is paid to syntax is that structural aspects of PROPS which transcend the word boundary are generally neglected because they do not seem to shed much light on the (cultural) history of PROPS which, however, is at the heart of the onomastic research programme (Eichler 1995: 1). Interestingly, the above authors only raise the question what linguistics can contribute to the study of PROPS but do not consider the possibility that an in-depth inquiry into the structural properties of PROPS might hold in store for linguistics. We are confident that our study will prove that the relation is not as one-sided as the HSK *Namengrammatik* seems to suggest.

Twenty-one years after the publication of the HSK *Namengrammatik* Hough (2016) edited the *Oxford Handbook of Names and Naming*. Of the forty-seven chapters therein, only one co-authored by Van Langendonck and Van de Velde (2016) deals with grammatical properties of PROPS albeit – despite the chapter's title – only in passing (Van Langendonck and Van de Velde 2016: 22–26). A handful of sentences from Dutch, Kirundi, and Namia complement the mostly English exam-

---

[22] Our translation: "since names are first of all linguistic signs, linguistics is the discipline whose responsibility it is first and foremost to study names."

ples. Since more space is reserved for the problems raised by definiteness, referentiality, semantics, and pragmatics of PROPs, the progress that has been made in the research dedicated to the grammar of PROPs since 1995 is hardly visible from this necessarily sketchy text which also contains an abridged taxonomy of PROPs which the authors classify as a typology (cf. below). In the conclusions, Van Langendonck and Van de Velde (2016: 38) claim that PROPs arguably are "the most prototypical nominal category." We will come back to this claim below. They also assume that

> the pragmatic-semantic concept of names [. . .] is cross-linguistically applicable. It is distinct from language-specific categories of Proper Names [. . .]. The language specific question as to what belongs to the grammatical category of Names does not necessarily yield the same answer as the question of what can be considered a name from the semantic-pragmatic point of view. Mismatches are most likely to be found at the bottom of the cline of nameworthiness [. . .]. (Van Langendonck and Van de Velde 2016: 38)

This quote calls to mind the extended PROP-N classes introduced in Section 1.2. We agree with the above authors that the grammar of the individual language determines what is and what is not a PROP. However, there is no absolute arbitrariness involved because it is possible to identify cross-linguistic similarities which are not mentioned as such in the reviewed texts. In sum, the grammar of PROPs is occasionally referred to in the extant onomastic literature but the topic is only of minor importance to the discipline. The apparent lack of interest on the part of onomastics should not be taken to mean that there is nothing worth talking about in the domain of the grammar of PROPs. It is the task of linguistics to demonstrate that PROPs are a promising research object for (comparative) grammarians.

2.2 The linguistic turn

In the first decade of the new millennium, things began to change. In Anderson (2004: 470), it is hypothesised – on the basis of comparative data from English, French, Greek, and Hungarian (but with assumed further generalisability) – that PROPs together with pronouns[23] and determiners constitute the class of determinatives, i.e., they have to be separated from the word-class of nouns. This idea is further elaborated upon within the framework of Notional Grammar (Anderson 1991) in the book-length presentation of the grammar of PROPs (Anderson 2007).

---

23 Throughout this study, the reader will find a number of languages which attest to identical morphological properties of PROPs and pronouns (in contrast to COMMs). We do not look into this issue systematically but are aware of the necessity that the parallel behaviour of PROPs and pronouns is a phenomenon worth investigating in-depth in a dedicated study of its own.

Additional data from a variety of further languages broaden the empirical basis of the approach without giving the study the character of a genuine typologically-minded cross-linguistic investigation.²⁴ Part III of the monograph addresses a selection of grammatical issues among which syntactic phenomena also play a prominent role (Anderson 2007: 239–286). Definiteness, genericity, reference, determination are recurrent themes throughout the better part of the book. Anderson's work is firmly situated in linguistics since

(a) the typical topics of onomastics are not taken account of and
(b) PROPs are analysed by way of applying the principles of a theory of grammar, namely Notional Grammar.

Van Langendonck (2007) contends against Anderson's central hypothesis according to which PROPs are dissociated from nouns. *Theory and Typology of Proper Names* can be understood as response to and rebuttal of Anderson's model. It is therefore hardly surprising to see that Van Langendonck (2007) comments on most of those themes which are featured in Anderson (2004, 2007), viz. definiteness, genericity, reference, and determination. In contrast to Anderson (2007) who surveys some grammatical properties of the grammar of PROPs to lend credence to his idea of subsuming PROPs, pronouns, and determiners under one umbrella, Van Langendonck (2007: 120–124) argues for classifying PROPs as a nominal category. He concludes that "proper nouns are more akin to common nouns than to pronouns" (Van Langendonck 2007: 171).²⁵ On certain parameters, PROPs tend to be unmarked which entails that the grammatical categories are frequently zero-coded²⁶ so that PROPs "are the most prototypical nominal category" (Van Langendonck 2007: 173) – a strong assumption to which the proponent still sticks nine years later (cf. above). He shares this opinion with other authors who do not specifically study the grammar of PROPs (Creissels 2006: 37). If, one might ask, PROPs are at the very heart of the system of nouns, can it be that anything like SOG exists? Would it not be the other way around? PROPs define how nouns have to behave – and if COMMs do not obey these rules, they give evidence of *Special Common Noun Grammar*... We

---

24 Several of the additionally integrated languages such as Albanian, Basque, Macedonian, etc. only serve to illustrate principles of Notional Grammar whereas their grammar of PROPs as such is not analysed at all. This must be borne in mind when one reflects upon the putative increase of cross-linguistic knowledge in the domain we are interested in.
25 Similarly, Croft (2022: 66) argues that "proper nouns [...] are a subtype of noun – that is, a word that denotes an object in reference."
26 In this study, we employ the traditional and well-established terms zero-coding and zero-marking for phenomena which in different contexts might better be labelled *Absence of Material Exponence (AOME)* as proposed by Stolz and Levkovych (2019b).

refrain from going on splitting hairs like this. What we learn from the prototypicality argument, however, is that the interpretation of the concept of *Special Grammar* depends almost entirely on one's perspective (cf. Section 2.3).

Van Langendonck's book comprises a detailed description of ANTHs in Flemish (Van Langendonck 2007: 256–320). The term typology – as already mentioned in Section 2.1 – is employed not in the sense of cross-linguistic research but as an alternative to the term taxonomy which comprises the subclasses of PROPs (Van Langendonck 2007: 183–253). Apart from the Flemish data, we find Dutch, English, French, and Polish examples alongside the occasional piece of evidence from extra-European languages. Van Langendonck (2007: 13–14) finds support for his own research in *Radical Construction Grammar* (Croft 2001) which means that he approaches the grammar of PROPs from a properly linguistic point of view.

It is beyond question that both Anderson's and Van Langendonck's studies have been decisive for the subsequently increased interest of linguists in PROPs. From 2007 onwards, a growing number of publications have appeared in which the linguistic side of PROPs is explored. Since the reviewed authors used empirical data mainly if not exclusively to substantiate their opposing standpoints – PROPs being or not being nouns – many gaps in the phenomenology of the grammar of PROPs remain to be filled. What is more, the increased number of languages which more or less extensively serve as illustrations of certain phenomena has not yet given rise to any serious cross-linguistic comparison of the grammar of PROPs which would satisfy the usual criteria of functional typology. Anderson and Van Langendonck have paved the way for future research from which the exhaustive description of the grammar of PROPs might emerge but they have not spoken the final word on this matter.

2.3 On being special

The first edition of Nübling et al. (2015) in 2012 added considerable zest to the study of the grammar of PROPs first in Germany and soon also internationally. We interpret the impressive list of pertinent publications prior and posterior to the above monograph – all of which go to the credit of Nübling and/or her associates – as a serious attempt to establish grammar in the canon of topics in onomastics. We ignore to what extent this laudable initiative has been successful in the onomastic mainstream. What must be conceded in any case, however, is that Nübling et al.'s projects have triggered a new research programme in which onomasticians and linguists co-operate (cf. Section 2.4). To keep the discussion of Nübling's properly linguistic approach to the grammar of PROPs within reasonable bounds, we pick out two aspects from her work which are especially interesting for our own project, namely the concept of SOG as such (= Section 2.3.1) and the pervasiveness of SOG in individual languages (= Section 2.3.2).

2.3.1 A plethora of special grammars: The case of Nungon

Nübling et al. (2015: 64) introduce the concept of SOG as follows:

> Namen befolgen eigene grammatische Regeln, die von denen der A[ppelativa] beträchtlich abweichen können. Wir sehen auch diese Abweichungen als funktional an: die **onymische Sondergrammatik** [original boldface] dient der Abgrenzung der E[igennamen] von den A[ppelativa], aber auch der Schonung ihres Namenköpers.[27]

It follows from this short description that SOG differs from the grammar of PROPS insofar as SOG focuses on the distinctive properties of PROPS which are not shared by COMMS whereas writing the grammar of PROPS means that all structural and functional properties of PROPS have to be accounted for no matter how widespread they are across other nouny word-classes and beyond. SOG has two functional motivations. First, it helps to draw a dividing line between subclasses of nouns. Second, the rules of SOG are such that their application allows the PROPS to preserve their phonological shape in order to guarantee easy recognition.

Nübling et al. (2015: 64–92) provide numerous examples of SOG phenomena, most of which show that PROPS and COMMS behave differently on a given parameter in German (and occasionally sundry languages). The diachronic dimension is especially highlighted since the authors are particularly interested in the dynamics which bring about changes from COMM to PROP and vice versa – the processes go by the names of deappelativisation (German: *Deappelativisierung*) and deonymicisation (German: *Deonymisierung*), respectively (Nübling et al. 2015: 49–63). In this way, it is shown that the boundary between PROPS and COMMS is permeable and that the dynamics which move one item from a given class to another are at work also synchronically. The processes are gradual, changes do not necessarily affect all members of a given class, at least temporarily, there might also be variation, etc. What all these provisos suggest is that the rules of SOG may frequently be only optional and limited to subsets of the entire onomasticon. In what follows, we will repeatedly have occasion to point out such cases to the reader. For the purpose of

---

27 Our translation: "Names follow their own grammatical rules which may considerably deviate from those of common nouns. We consider these deviations too as being functional: **Special Onymic Grammar** serves to separate proper names from common nouns, but also to preserve their segmental form."

this study, no attempt is made to systematically account for diachronic aspects associated with the grammar of PROPS.

The way SOG has been introduced above might give rise to the mistaken idea that there is only one special grammar. As a matter of fact, there is ample space for many special grammars to co-exist in one and the same language. The Papuan language Nungon (Northeast New Guinea) is described in Sarvasy (2017). The author of the grammar dedicates sixteen printed pages to the discussion of the word-class of nouns and its subclasses (Sarvasy 2017: 122–137) to which ANTH and TOPO belong (Sarvasy 2017: 132–135), too. There is no homogeneous COMM-category. As Table 4 reveals, there is a fine-grained differentiation of nouny words. The leftmost column hosts the original class ISO-code. There are three major classes A, B, and C, each of which comprises at least two subclasses. In the case of B3, there are even four sub-subclasses. The following colour code is used: yellow marks out those cells in which the same value as in ANTH **and** TOPO is given, grey shows where cells correspond only with TOPO, whereas blue identifies all cases of exclusive agreement with ANTH. Note that a third name class – song corpora names – is featured as C3 in Table 4. What strikes the eye most is the fact that both ANTH and TOPO fail to meet four out of five criteria. The one that ANTH fulfils is not the same as the one that TOPO fulfils. These name classes are thus mainly negatively defined, i.e., they are not allowed to do what other subclasses legitimately do.

Table 4: Nominal subclasses compared (Sarvasy 2017: 123).

subclass		number	pertensive	address	NP-head	LOC -*in*
A1	kin terms	yes	yes	yes	yes	no
A2	human	yes	yes	no	yes	no
B1	body part	no	yes	no	yes	some
B2	artifact component	no	yes	no	yes	some
B3	entire entity	no	no	yes	yes	some
B3a	generic, higher taxa	no	no	poetry	yes	no
B3b	specific, lower taxa	no	no	poetry	no	no
B3c	locational	no	no	no	yes	yes
B3d	temporal	no	no	no	yes	some
B4	non-specific	'who'	no	not alone	no	some
C1	ANTH	no	no	yes	no	no
C2	TOPO	no	no	no	no	some
C3	song corpora name	no	no	no	no	no

In Table 4, no two rows are filled identically. Accordingly, one could speak of thirteen different classes – or thirteen different special grammars, only three of which involve PROPS. It is cross-linguistically common that kinship terms and body-part terms behave differently from other COMMS and PROPS under possession where languages formally distinguish inalienable from alienable concepts (Chappell and McGregor 1995). There is thus the possibility of what might be called *Special Kinship Grammar* (Dahl and Koptjevskaja-Tamm 2001) and *Special Body-part Grammar* (Kukuczka 1982) to arise. Similarly, human nouns rank very high on the animacy/empathy hierarchy which privileges them ever so often also grammatically in the sense that certain categories are only expressed for them such as number in Classical Nahuatl (Launey 2011: 19–22). There are thus good grounds for conceiving the *Special Human Grammar*, too. This list could easily be extended by further candidates. Note that if PROPS – presumably only ANTHS – were made the point of departure for determining what is Special Grammar in the domain under scrutiny in accordance with the prototypicality assumption mentioned in Section 2.2, all but one of the classes in Table 4 would host Special Grammars. It is obvious that SOG is not unique; it is only one of an undetermined number of other special grammars, the domains of which might at times overlap partly.

The differences between the classes in Table 4 can be minimal. Between C2 TOPO and B3d temporal nouns, there is just one cell which does not host the same value for both classes. Similarly, C1 ANTH and B3b specific/lower taxa differ only on a single parameter. The situation captured by Table 4 resembles a continuum on which it is difficult to locate cut-off points between categories. Accordingly, if the label SOG is used to characterise a given phenomenon, this does not automatically entail that the linguist is facing sets of major distinctive properties. There are many cases which can be classified as micro-level dissimilarities between PROPS and COMMS. Moreover, Nübling et al. (2015: 65) refer to Harweg (1999: 195) who, in speaking only about the marking of PROPS, can be interpreted to mean that if there is evidence of SOG the property might not count for all PROPS but only for a certain subset thereof. Table 4 tells us that no ANTH hosts the locative suffix *-in* whereas some (but not all) TOPOS do (Sarvasy 2017: 408). In this way not only ANTHS are distinguished from TOPOS but within the TOPO-subclass there is also variation among its members as shown in (20).[28]

(20) Nungon [Nuclear Trans New Guinea, Finisterre-Huon]
 (a) without suffix (Sarvasy 2017: 414)
  Ongo-ng-a   Worin-Ø   ir=it-do-mong
  go-SUBJ-MEDV Worin-Ø   be=be-REMPST-1PL
  'Going along, we used to stay **in** Worin.' [O.T.]

---

[28] The TOPOS in (20) and the next paragraph refer to settlements in Northeast Newguinea in the vicinity of Towet, the village where the data for the Nungon grammar were collected.

(b) with suffix (Sarvasy 2017: 541)
Gaga=wuk   höm   Yapem-**in**   hai-wi-rok-ma
1SG.EMPH=RFL houspost   Yapem-LOC   fell-IRR.SG-2SG-FUT
yo-ng   Gaus   gogo=gon
say-SUBJ   Gaus   1SG.FOC=RSTR
'"By yourself, the houseposts, you will fell them **at** Yapem," saying, "Gaus, just you will."' [O.T.]

The TOPOS *Worin* and *Yapem* display morphological differences in the sense that for the expression of an identical function (here the spatial relation Place), *Worin* remains uninflected whereas *Yapem* takes the locative suffix *-in*. Since TOPOS like *Towet, Yawan, Köngsuenon, Inabö, Komutuk*, etc. (Sarvasy 2017: 456–457) also lack the locative suffix when functioning as Ground in Goal, Place, and Source relations, the absence of *-in* in (20a) cannot solely be explained by haplology, i.e., the avoidance of the sequence of *-in-in*. As far as we can tell, the blockage of locative inflection on PROPS has not been addressed in the literature mentioned in Sections 2.1–2.2. The class of TOPOS is divided in two by the criterion of *in*-suffixation. This is tantamount to the existence of two **inflectional classes** which, in turn, count among the morphological mismatches catalogued in Corbett (2007: 30).

Table 4 shows that neither ANTHS nor TOPOS are subject to number marking (Nungon having singular, dual, and plural number). The absence of number distinctions is unsurprising since Anderson (2007: 234) and Nübling et al. (2015: 72–73) agree on the limited possibilities of pluralising PROPS in the languages they study. Van Langendonck (2007: 159) emphasises that there are PROPS which can undergo pluralisation like the average COMM. In Nungon, PROPS cannot be NP-heads – a negative property which implies that adjectival attribution is blocked. Dixon (2010a: 108) assumes that generally

> [a] proper noun as the head of an NP is likely to have far fewer – if any – possibilities for modification, when compared to a common noun as NP head. It may be the case that it does not accept an adjective, nor a relative clause, nor a demonstrative.

In contrast, Van Langendonck (2007: 121) insists that phrases "whose head is a proper name have the external and internal properties of NPs." To prove his point, he presents examples from Dutch and English which involve adjectival attribution of PROPS (cf. Handschuh 2022: 41). The inability of Nungon PROPS to inflect for the pertensive has no counterpart in any of the linguistic studies mentioned in the foregoing sections – and not only because the term pertensive was coined by Dixon (2010b: 268–312) only relatively recently. Simplifying, the pertensive applies if in an adnominal possessive construction, the possessee bears the marker of the relation to the possessor (Sarvasy 2017: 428–429). Moreover, an important category is missing from Table 4 because Sarvasy (2017) did not include it in the original, namely the associative plural. This category is restricted to ANTHs because

> the associative plural constructions [...] use the name or epithet of a particular person or people as the label for a group associated with that person or people. (Sarvasy 2017: 177)

Of the three different types of associative plural, we only illustrate the so-called general pertensive associative plural with an example from the reference grammar of Nungon. In (21), we find the realisation of the construction pattern
– [ANTH N$_{KINSHIP}$-*nit*]$_{ASSPL}$.

To detect constructions of this kind one has to look beyond the word boundary.

(21) Nungon [Nuclear Trans New Guinea, Finisterre-Huon] (Sarvasy 2017: 178)
 Gorungon nip-nit e-wa-ng
 Gorungon cross_cousin-ASSPL come-PRS.NSG-2/3.PL
 'Gorungon **with** his cousins are coming.'

Sarvasy (2017: 178) states that "[t]he personal name or epithet that serves to label the group is always that of an individual." To our mind, this is an important property which deserved a column of its own in Table 4 because it singles out ANTH from all other nouny classes. In their chapter on associative plurals in the WALS, Daniel and Moravcsik (2005: 150) state that "associative plurals are much more typical of proper names than of common nouns." Mauri and Sansò (2019: 622) narrow the scope down to "a construction type that applies to proper names and kin terms" in numerous languages. Aikhenvald (2015: 113) goes a step further when she claims that

> [t]he associative plural, and the associative dual [...] constitute a useful criterion for personal names, or kin names as a separate subclass of nouns [...].

The author indirectly declares the associative numbers to be an important piece of evidence for the existence of SOG.

Nungon is thus no outsider since associative plurals of ANTHs are cross-linguistically relatively frequent (cf. Iloko in Section 1.2) which, in turn, requires that it is accounted for in the morphosyntactic typology of PROPs (Handschuh 2022: 32–33). Glossing over the fact that a homogenous COMM-class cannot be identified, we may still assume that Nungon realises the formula COMM ≠ TOPO ≠ ANTH.

Typical categories like definiteness which have been keeping scholars busy in the domain of the grammar of PROPs do not play any discernible role for SOG in Nungon. In contrast, other parameters such as locative marking and associative plurals are crucial for SOG in this language. We conclude that there is an urgent need for complementing the phenomenology of SOG by way of looking (far) beyond the canon of topics that have dominated the debate ever since. To achieve this, a typological stance is required of the linguists who investigate into the nature of SOG. The previous empirical basis with its strong (Indo-)European bias has proved to be too small to give us the full picture of how SOG can look like.

2.3.2 *Special Onymic Grammar* all over the place: Vurës

A look beyond the confines of the languages with which the linguist is particularly familiar is definitely called for when hypotheses are put forward for which universal validity is assumed. Nübling et al. (2015: 64) for instance speculate about the theoretical possibility of the existence of a language which consistently marks COMMS and PROPS throughout the lexicon and the onomasticon. In passing and with reference to Seiler (1983a: 154), they mention Tagalog where this distinction supposedly (as they cautiously add) is attested. Parallels in Fijian and Hua are mentioned too with reference to Hockett (1958: 311–312) and Löbel (2002: 592), respectively. Our case study on Iloko in Section 1.2 can be added to this short list – and many more languages too.

Within the context of an in-depth study of SOG in an individual language, sideways remarks on other languages like the above are certainly tolerable. They are, however, insufficient and inadequate if one aims at extrapolating from one's data over ideally all languages. There is also the danger of exoticising those languages which are touched upon in passing only. To demonstrate that there might hide much more behind the consistent overt marking of PROPS vs COMMS, we have a closer look at another member of the Austronesian language family.

For Vurës – an Oceanic language spoken in Vanuatu – Malau (2016: 102) postulates a layered system of subclasses of nouns which we represent in Figure 1 as a tree diagram making use of the terms and abbreviations introduced in Section 1.2.[29]

Kinship terms which can be used as forms of address without further morphological manipulation are counted as ANTH whereas those kinship terms which are obligatorily possessed fall under ANTH-N. Names of tribes form the class of TOPO-NS (Malau 2016: 103–105). The grammarian of Vurës provides detailed checklists in which the different morphosyntactic behaviour of ANTH(-N), TOPO(-N), and COMM is spelled out. Table 5 synoptically bundles this information which Malau (2016: 102, 106–107) distributes over three different paragraphs.

---

[29] Malau's (2016) own terms are *proper noun* for our ANTH(-N), *free personal name* for ANTH, and *absolute location nouns* for TOPO(-N).

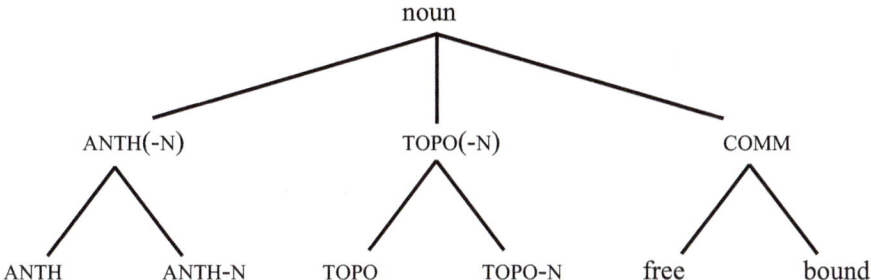

Figure 1: Subclasses of nouns in Vurës.

Table 5: Properties of subclasses of nouns in Vurës.

criterion	COMM	ANTH(-N)	TOPO(-N)
article	o na / POSS	i [PL ira] na [ANTH-N_POSS]	a [– ta __]
interrogative	so	sē	vē
demonstratives	yes (all)	yes (certain)	e
IO=clitic	yes	no	no
prepositions	most	most	some
relative clause	yes	yes	non-restrictive
adjectives	yes	(yes)	no
numerals	yes	yes	no
[NP-head]_argument	yes	yes	no

The grey-shaded cells identify those parameters on which TOPO(-N)s behave differently from COMM. The bluish colour tells us that there is a behavioural difference between ANTH(-N) and COMM. The zone of overlap of COMM and TOPO(-N) is empty whereas ANTH(-N) shares six properties with COMM. Only once do ANTH(-N) and TOPO(-N) show identical behaviour, namely in the case of the impossibility to host the IO=clitic[30] (cf. below). This means that if both ANTH(-N) and TOPO(-N) disagree with COMM on a given parameter they usually do it in different ways. Some of the differences are morphological as, e.g., when COMMs are accompanied by the article o but ANTH and TOPO take their articles i and a, respectively (cf. (22)).[31]

---

[30] In this and other cases, the capital <O> is a variable for a vowel whose quality changes according to the rules of vowel-harmony in Vurës.
[31] If used as glosses, the abbreviations ANTH, COMM, PROP, and TOPO denote markers specific to the respective categories as, e.g., in (22a–c).

(22) Vurës [Austronesian, Oceanic]
    (a) COMM (Malau 2016: 158)
        ni   ōn-veg   o   ev
        3SG.DEF   lie-APPL COMM fire
        '[. . .] he lay by **the** fire.' [O.T.]
    (b) ANTH (Malau 2016: 238)
        i   *Catriona* *ni*   *sum*   *gē*
        ANTH   Catriona 3SG.DEF drink kava
        '[. . .] Catriona drank kava [. . .].' [O.T.]
    (c) TOPO (Malau 2016: 224)
        Rōrō *a*   *M̄ikian*
        3DU   TOPO   M̄ikian
        'The two of them were at M̄ikian.' [O.T.]

The overt distinction of three classes by way of different pre-nominal morphemes is reminiscent of the situation in Iloko as described in Section 1.2. This fact alone would justify considering Vurës to realise the formula COMM ≠ TOPO ≠ ANTH. Differences of this kind are overt in the sense that they involve phonologically realised markers. On other parameters morphosyntactic and syntactic factors come into play such as the combinability with modifiers, determiners, numerals, or prepositions and the ability of functioning as NP-head in argument function, etc. These differences are mostly covert because they affect the combinatory potential and/or the syntactic functions of the nouny items. The covert nature of many SOG-phenomena renders the search for proof of SOG particularly demanding since these pieces of evidence are easy to overlook – and Malau's (2016) grammar of Vurës is certainly exceptional because of the meticulous care given to the distinction of PROPS and COMMS in this language. Note also that Van Langendonck and Van de Velde's (2016: 23) claim that PROPS cannot be modified by restrictive relative clauses is met only by TOPOS in Vurës. In (23), we present examples of relative clauses whose head is an ANTH.

(23) Vurës [Austronesian, Oceanic]
    (a) restrictive relative clause (Malau 2016: 664)
        ine   *[i*   *Qet*   *[so*   *ta=qa~qaq*   *min*   *i*   *Dōl]$_{RRC}$]$_{NP}$
        DEM [ANTH Qet [REL PROG=DUR~talk DAT ANTH Dōl]]
        'That is Qet **who** was talking to Dōl.' [O.T.]
        'That is the Qet **that** was talking to Dōl.' [rephrased]

(b) non-restrictive relative clause (Malau 2016: 666)

nēk	i	kal	liñereg	lö=lölö	mōtō	bule-n
2SG	2SG.DEF	go_in	inside	COM.LOC=inside	coconut	CL.DOM-CONST

[Hosea Waras [Ø ta=kakaka lö=lölö kaset ko]_{NRRC}]_{NP}
Hosea Waras Ø PROG=speak COM.LOC=inside cassette PRX

'[...] you go into the coconut plantation of Hosea Waras **who** is speaking on this cassette [...].'

Malau (2016: 658–667) provides a detailed account of relative clauses in Vurës. The relativiser *so* is optional so that "the intonation is the only indication that the relative clause is part of the NP and not a separate clause" (Malau 2016: 660). The presence of *so* in (23a) as opposed to its absence in (23b) is no correlate of the distinction between restrictive and non-restrictive relative clauses. The possibility of ANTHS functioning as heads of restrictive relative clauses in Vurës seriously challenges the probably Eurocentric hypothesis according to which PROPS are generally excluded from modification by restrictive relative clauses.

Vurës has still more on offer. As to ANTHS, there is the "female personal name prefix *rO-* [which] only occurs on personal names of females and can be optional, particularly with non-indigenous names." (Malau 2016: 159). The diachronically corresponding masculine ANTH prefix *wO-* has undergone generalisation to attain the status of a "general default noun prefix" (Malau 2016: 159). Since Vurës is a language without grammatical gender, we are dealing with female sex marking as a derivational means (because most of the traditional ANTHS are derived from COMMS) whereas imported ANTHS are underived and thus may function without the prefix (Malau 2016: 159–160). The prefix *dO-* is an interesting parallel. It can only be attached to COMMS referring to plants in order to derive expressions which refer to the leaves of a given plant.

Outside the domain of derivational morphology, we find the *lO*=clitic which is compatible only with COMMS whereas it is banned from combinations with PROPS. Malau (2016: 239) considers *lO=* to be

> the only article that has case marking properties. It only occurs when the head noun is a common noun [...] and specifies that the NP is a locative adjunct. This article is thus in complementary distribution with *o* [...].

Accordingly, *lO=* is labelled as "common locative article" as opposed to *a*, the "absolute locative article" (Malau 2016: 240). Two comments are in order at this point.

(i) The inability of PROPS to combine with a certain marker of spatial relations can be compared to the differentiation of forms for the oblique in Iloko (Section 1.2) and the blockage of the locative suffix *-in* with ANTHS and some TOPOS in Nungon (Section 2.3.1).

(ii) Malau (2016: 223) emphasises that TOPO(-N)s may "form the head of an NP, however, the NP can only be a clausal adjunct [...], or a predicate of a non-verbal clause [...], not an argument of the clause." Thus, *a* does not only mark the class-membership of a TOPO but also its syntactic function as adjunct.

As to (i), one is reminded of a remark in Nübling et al. (2015: 86) where the authors mention the use of special prepositions and the special use of prepositions in PPs whose complement is a PROP in German. For Nübling et al. (2015: 86) this phenomenon belongs peripherally to the domain of syntax. We add that one has to account for the semantic domain as well because like the cases mentioned in (i), the German prepositions to which Nübling et al. (2015) allude express certain spatial relations which cross-linguistically form part of the centrepiece of SOG as will come to the fore in Section 5. The absence of TOPOs from the argument functions belongs to the "general syntactic rules [which] dictate that a morphosyntactic value is not relevant for some lexeme" (Baerman and Corbett 2010: 3), in our case the irrelevance holds for an entire class of nouns. Within the macro-class of nouns, TOPOs instantiate **defectiveness**, i.e., a deviation from the morphological canon.[32]

The exclusion of TOPOs from the syntactic functions of subject and object claimed in (ii) has so far largely escaped being noticed by linguists with an interest in the grammar of names, in general and SOG, in particular. A distant parallel in Classical Nahuatl is described in Stolz et al. (2017a: 140–142) (cf. below). Since both COMMS and ANTHS are licit as arguments of predicates in Vurës, TOPOs behave markedly differently from the majority of nouns. One might want to argue that not all PROPS are prototypical nouns if one of their classes is not allowed to fulfil the basic syntactic functions which are, however, open to COMMS.

The case of Vurës confirms what those of Iloko and Nungon have already shown, namely that no matter how detailed the extant catalogues of SOG-phenomena are – be they dedicated to individual languages such as German (Nübling et al. 2015) or meant to have cross-linguistic scope (Anderson 2007, Van Langendonck 2007, Van Langendonck and Van de Velde 2016) – there are far too many properties of PROPS which have not yet been accounted for (because they have not yet been discovered!) to declare the case closed. We have only just begun to venture into still largely uncharted territory.

---

[32] Caution is called for in connection with this hypothesis. The reference grammar of Vurës does not tell us whether and how it is possible to predicate over TOPOs. At the same time, ANTHS are probably also defective because their failure to take the *lO*=clitic might mean that ANTHS cannot function as Ground in spatial situations. No alternative strategy of marking ANTHS in this function are mentioned in Malau (2016).

2.4 Broadening the horizon

2.4.1 Beyond German (with glimpses of Palikur and Dargwa)

The merits of Nübling and associates are great. It is their initiative which is the point of departure for a new wave of studies in the domain of the grammar of PROPs. The only criticism that could be directed to their approach also holds for those of Anderson and Van Langendonck, namely that the sample of languages, on the properties of which generalisations about the grammar of PROPs are based, is far too small and too biased to extrapolate from. Admittedly, the criticism is unfair in the case of Nübling et al. because their primary goal has been to determine SOG for German and all its varieties past and present. Numerous publications on SOG phenomena in German and other Germanic languages – very often with a diachronic orientation have sprung from the work of this research team so that the knowledge about the grammar of PROPs and SOG in the Germanic phylum has increased considerably. Since our own focus is on typological matters and cross-linguistic data, we mention only the papers by Berchthold and Dammel (2014) and Nübling (2017a, b) out of a plethora of other studies. Definiteness marking, the tendency towards the formal dissociation of PROPs from COMMs, and the assignment of PROPs to grammatical genders are recurrent themes in this strand of research. Especially the first mentioned one is also prominent in many studies devoted to differently affiliated languages.

Inspired by the work of the research team at the *Johannes-Gutenberg-University* and the *Akademie der Wissenschaften und Literatur* (both located in Mainz), linguists specialising in languages other than the Germanic ones have begun to look into the grammar of PROPs of their object languages. The following (most probably incomplete) list of publications is chronologically ordered (not all of them are directly attributable to the influence of the centre of diffusion of ideas in Mainz):

- Szczepaniak (2005) presents data from Polish (Slavic) morphology which are suggestive of the emergence of a PROP-marking strategy.
- Mojapelo (2009) discusses semantic and morphological properties of ANTHs in North Sotho (Bantu).
- Van de Velde and Ambouroue (2011) provide a survey of the grammar of PROPs in Orungu (Bantu).
- Caro Reina (2014) traces the emergence of ANTH articles from terms of address in Balearic Catalan (Romance).
- Stolz et al. (2014: 188–224) describe how a subclass of TOPOs – hodonyms and dromonyms – systematically alternate between overt marking and zero-marking of spatial relations in French (Romance).

- Le Bihan (2015) compares the Celtic languages Welsh and Breton as to the syntactic status of TOPOS.
- Stolz et al. (2017b) take zero-marking of spatial relations as evidence of SOG in Maltese (Afro-Asiatic).
- Heidenkummer and Helmbrecht (2017) discuss the genesis of a dedicated ANTH marker in the Siouan language Hoocąk.
- Stolz and Levkovych (2019a) highlight aspects of SOG in the domain of TOPOS in Aromanian (Romance).
- Boyeldieu (2019) shows that the choice of case markers differs between COMMS and ANTHS in the Chadic language Sinyar.
- Caro Reina and Nowak (2019) describe the diachronic developments in the gender assignment of city names (= a subclass of TOPOS) in Spanish (Romance).
- Stolz (2019) dedicates his study to the blockage of initial mutations with PROPS in Welsh (Celtic).
- Helmbrecht (2020b) sketches the morphosyntax of ANTHS in Hoocąk (Sioux).
- Caro Reina (2020a) investigates the employment of definite articles with TOPOS in the Romance phylum.
- Caro Reina (2020b) determines to what extent ANTHS are subject to differential object marking in Romance languages.
- Stolz (2020) is an in-depth study of the variation in the use of the TOPO-article in the Austronesian language Chamorro.
- Stolz and Levkovych (2020a) review examples of languages whose TOPOS integrate spatial case markers, the role of which can be argued to be ambiguous between derivation and inflection.
- Stolz and Levkovych (2020b) investigate the similarities and dissimilarities of classifier types in the domains of TOPOS and COMMS.
- Caro Reina (2022) provides a detailed synchronic and diachronic account of the use of definite articles with ANTHS in Romance languages.
- D'hulst et al. (2022) pick out a special subclass of TOPOS – hydronyms – to compare West Germanic and Romance languages with reference to the use of definite articles with hydronyms.
- Jeffay and Rothstein (2022) focus on the behaviour of ANTHS in the construct state of Modern and Biblical Hebrew (Afro-Asiatic).
- Salaberri (2022) inquires into the possibilities of definiteness marking in the synchrony and diachrony of ANTHS in the isolate Basque.
- Stolz and Levkovych (2022) account for definiteness marking on PROPS in Fijian with comparative data from additional Austronesian languages.
- Nintemann and Hober (in press) investigate the encoding of Goal and Source with TOPOS as compared to COMMS across a wide range of Pidgins and Creoles.

This list features studies dedicated to individual languages from different branches of the Indo-European language family (Celtic, Romance, Slavic), Afro-Asiatic languages (Maltese and Hebrew), Austronesian languages (Chamorro and Fijian), a representative of Siouan, another from Chadic, and several Bantu languages as well as Pidgins and Creoles. Several studies are comparative across a given phylum or even transgressing boundaries between phyla.

The case studies in Sections 1.2, 2.3.1, and 2.3.2 show that it is relatively easy to add further languages to the inventory of those languages whose grammar of PROPS has been subject at least once in a linguistic publication. It goes without saying that there still are huge gaps in the coverage of languages. Similarly, there is a clear preference for a small selection of preferred topics, too, with definiteness standing out. A third of the twenty-four studies enumerated above address definiteness marking or articles tout court. Expectedly, this relatively oft-studied phenomenon also invites being studied typologically by way of comparing languages cross-linguistically (Helmbrecht 2022), cf. below. However, there are many more topics which could be addressed in this way, provided more languages are taken on board and along with them also new themes. We try to prove this by way of comparing two languages which so far have not been part of the programme of SOG studies.

In the Arawakan language Palikur, TOPOS have a special status. Launey (2003: 139) notes that case inflection in Palikur is doubly restricted, namely
(a) only spatial relations are expressed via suffixes and
(b) only TOPOS (and some adverbs) are inflected for these categories.

There is a small class of TOPO-NS as well like *paytwempu* 'village' and *parahwoka* 'sea' which can optionally take the same case-inflections (Launey 2003: 140). Four cases are distinguished formally: LOCATIVE (zero-marked) ≠ ALLATIVE (*-(V)t*) ≠ ABLATIVE (*-(V)tak*) ≠ PERLATIVE (*-(V)w*). Examples are given in (24).

(24) Palikur [Arawakan] (Launey 2003: 139–140)
   (a) locative
      Ig   ***ay***   *Kayan-Ø*
      3SG.M   **here**   Cayenne-Ø
      'He is (**here**) at Cayenne.'
   (b) allative 205
      Ig   *atak*   *Kayan-**it**?*
      3SG.M   go   Cayenne-**ALL**
      'He goes **to** Cayenne?'

(c) ablative
   Ig   ayta   Kayan-*itak*
   3SG.M   come   Cayenne-ABL
   'He comes **from** Cayenne.'
(d) perlative
   Ig   mpiye   Kayani-*w*
   3SG.M   pass   Cayenne-PERL
   'He passed **through** Cayenne.'

To express these spatial relations with COMMS, prepositions have to be used such as *-giku* 'inside' in (25).

(25) Palikur [Arawakan] (Launey 2003: 149)
   Kuruku a-giku   miyokwiye
   rat   POSS.NT-**inside** hole
   'The rat is **in** the hole.'

There are numerous prepositions, most of which serve to express specific location whereas the case inflections on TOPOS are examples of general location. The distinction of general and specific location becomes relevant again in Section 5 where definitions are provided. Examples of ANTH functioning as Ground are not provided in the grammar. This data gap notwithstanding, we assume that Palikur realises the by now familiar formula COMM ≠ TOPO ≠ ANTH. The morphological mismatch we are faced with here is **anti-periphrasis** (Haspelmath 2000: 659), i.e., TOPOs use inflected word-forms where other nouns employ two-word constructions (= periphrases)

More importantly, Launey (2003: 140) states that

> [c]omme le locative n'a pas de marque spécifique, un toponyme marquant une localisation sans movement est en générale accompagné d'un adverbe come **ay**, *ici* ou **ayhte**, *là-bas* (pour éviter d'être interprété comme sujet ou objet) [original boldface and italics][33]

Whether the functional reason given in brackets by the quoted author is valid does not concern us here. For the point we are going to make it is more important to recognise the similarity between Palikur and Iloko (Section 1.2). Both languages resort to deictic elements to express certain spatial relations with TOPOs. In both languages, the choice of the appropriate deictic depends on the whereabouts of

---

33 Our translation: "since the locative has no specific marker, a place name marking a localisation without motion is generally accompanied by an adverb like *ay* 'here' or *ayhte* 'there' (to avoid being interpreted as subject or object)."

the speaker. Since the two languages belong to different language families and are spoken on different continents, the structural similarities are independent of each other. They are suggestive of general principles which determine how the morphosyntax of spatial relations can be shaped in the domain of TOPOs. This possibility must escape one's notice if only individual languages are studied separately. We hypothesise that there are many more languages at different places on the world map which attest to similar strategies of employing deictics in the paradigm(s) of TOPOs.

In Palikur, only TOPOs display case inflection. In the Nakh-Daghestanian language Sanzhi Dargwa, "place names only inflect for directional cases (essive, lative, ablative)" (Forker 2020: 198) whereas COMMs boast four grammatical and nineteen semantic cases (Forker 2020: 54). Since there is also evidence of TOPOs taking the genitive suffix the above constraint probably needs to be revised.[34] Forker (2020: 198) further remarks that the basic form of a given TOPO usually has directional meaning so that the allative is zero-marked. Zero-marking of Place seems to be possible also with TOPOs as Grounds. The full paradigm with overt case marking as, e.g., for the TOPO *uc'ari* 'Itsari' is

- LOC *uc'ari-b* 'in Itsari' ≠ ALL *uc'ari-Ø* 'to Itsari' ≠ ABL *uc'ari-rka* 'from Itsari'.

Sentential examples involving two instances of zero-marking are given in (26).

(26) Sanzhi Dargwa [Nakh-Daghestanian, Dargwic]
  (a) locative (Forker 2020: 201)
     han   b-irk-u   ix-t:i   š:ik'e-**b**
     remember HUM.PL-OCCUR.IPFV-PRS DEM.up-PL Shike-ESS
     'As I remember, they were **in** Shike.' [O.T.]
  (b) allative (Forker 2020: 199)
     du   priziw-li   ka-Ø-ač'-ib=da   urkaraqari-**Ø**
     1SG call-ERG   down-M-come.PFV-PST=1 Urkarakh-Ø
     'I came **to** Urkarakh by call.' [O.T.]
  (c) ablative (Forker 2020: 199)
     ca   uc'ari-**rka**   ca   χudec'a-**rka**   ca   šaʕrʔaʕ-**rka**
     one Itsari-ABL   one Khuduc-ABL one Shari-ABL
     '[...] one **from** Itsari, one **from** Khuduc, one **from** Shari.' [O.T.]

The frequently used Russian TOPOs which are not invested with an inherent locative meaning for Sanzhi Dargwa speakers normally require the overt marking

---

[34] See for instance *uc'ri-la š:i* {Itsari}-{GEN} {village} 'the village of Itsari' (Forker 2020: 199).

of spatial relations also for those relations which might be left unexpressed with genuinely Shanzi Dargwa TOPOS (Forker 2020: 199). In sum, Shanzi Dargwa TOPOS display defective paradigms when compared to those of COMMS and ANTHS. Since the latter two seem to share identical paradigms, this is a case of (COMM = ANTH) ≠ TOPO, i.e., an instance of *Special Toponymic Grammar (STG)* which we look into further in Section 5.

What is especially intriguing about the Shanzi Dargwa case is that it fits in perfectly with the previous one from Palikur although the structural particulars are widely different on the micro-level. The two cases instantiate the same in different ways. In both languages, TOPOS are singled out from the macro-class of nouns insofar as they behave differently from other nouny classes in the functional domain of the grammar of space. TOPOS are special exactly when it comes to expressing spatial relations. Similar observations have already been made in connection to Iloko (Section 1.2), Nungon (Section 2.3.1), and Vurës (Section 2.3.2). TOPOS seem to be especially prone to SOG-behaviour when pressed into service for the expression of spatial relations. Since this is a pattern that is attested in a variety of genetically, areally, and structurally different languages, it seems to be a linguistically promising project to study this and related phenomena cross-linguistically. This is the task of typologists.

2.4.2 Where's typology?

Karnowski and Pafel (2005) ask how different PROPS are from COMMS. The answer to this question still needs to be given. Kempf et al. (2020: 7) are certainly right when they complain that "[*d*]*as* [original italics] Überblickswerk zur spezifischen Grammatik von Eigennamen existiert bis heute nicht annähernd."[35] Typology is partly to blame for this situation because hitherto PROPS have been given too little attention in cross-linguistic research. In four widely used introductions to typology, there is either no mention of PROPS at all (Whaley 1997, Moravcsik 2013, Song 2018) or they are mentioned only in passing as in Velupillai (2012: 156, 158, 164, 255) without being considered a typological research object in their own right. PROPS are sometimes even explicitly excluded from the analysis as, e.g., in Rijkhoff's (2002: 19) research on the NP. In contrast, especially ANTHS pop up time and again when animacy-based differential choice of constructions is discussed (Croft 2022: 69–70). Contexts in which ANTHS feature prominently are for instance the accessibility scale

---

35 Our translation: "nothing like *the one* synopsis of the special grammar of proper names exists today."

of referring phrases (Croft 2022: 81), their role as object modifiers (Croft 2022: 141 and 154), and in preferred argument structure (Croft 2022: 249 and 260), to name only a few from a potentially very long list. Popping up does by no means imply that the role of ANTHs is discussed at length. TOPOs too are largely neglected in the general typological literature. Their fate is perhaps even worse than that of ANTHs.

Recent years, however, have witnessed the emergence of a new typologically-minded strand of research in the domain of the grammar of PROPs. The research team headed by Johannes Helmbrecht at the Institute of Linguistics at the University of Regensburg is responsible for a series of studies which reflect the progress of their cross-linguistic project on *Special Anthroponymic Grammar (SAG)*.

- Handschuh (2017) investigates SAG typologically in connection with definiteness and case.
- Helmbrecht et al. (2018) survey the morphosyntactic behaviour of ANTHs across a typologically sound sample of languages to determine to what extent SAG meets the expectations of the universal animacy hierarchy.
- Thanner (2019) takes up this issue by way of focusing on the morphosyntax of ANTHs in languages of the split-ergative type.
- Handschuh (2019) is a study of the patterns which arise when ANTHs are integrated into languages equipped with systems of gender, noun classes, etc.
- Helmbrecht (2020a) looks into the formal and functional diversity of ANTHs in the morphosyntax of a number of unrelated languages.
- Helmbrecht (2022) reports on the cross-linguistic results gained from a detailed study of the combinability of definite articles with ANTHs.
- Handschuh (2022) is the first typological survey of the structural differences which distinguish ANTHs from COMMs. Since this paper lays the foundations for a research programme dedicated to SAG, it deserves further comments.

Handschuh (2022) ticks off a long list of categories. In morphology, differences between ANTHs and COMMs (and unsystematically also TOPOs) are identified in the domains of definiteness, case, number, and gender. The author concludes for morphology that

> [w]hile the often-noted tendency of names to allow for less variation and to avoid overt marking that changes the phonological shape of the name [. . .] is present in a number of languages, the opposite can also be found. [. . .] Also, the claim that personal names are the category most likely to show distinct behaviour from common nouns has been challenged by [. . .] place names [. . .]. (Handschuh 2022: 38)

As to syntax, apposition, NP modifiers, and coordination are discussed. Handschuh (2022: 45–46) concludes that

> [p]roper names exhibit various grammatical differences from common nouns in the languages of the world. We are far from having a complete understanding how these differences are manifested. [...] The paper has also demonstrated that proper names do not necessarily exist as a monolithic category in the languages of the world. The two best-studied types of names – personal names and place names – can behave in a way that warrants assigning them to different parts-of-speech [...].

We fully subscribe to these statements. It is clear that the typological stock-taking and evaluation of the grammar of PROPS is still in its infancy. Many more studies written in the spirit of Handschuh (2022) are needed before we can seriously claim that we are sufficiently informed about PROPS to put forward a viable typologically-oriented theory of the grammar of PROPS.

One way of reaching this distant goal consists in covering ever more phenomena. In the subsequent paragraphs, we want to briefly present a gap that calls for being filled. Vocatives are mentioned as a category that is often the monopoly of ANTHS (Aikhenvald 2015: 89). Anderson (2007: 219–221, 279–286) devotes two subsections of his monograph to the relation between PROPS and vocatives. It seems, however, that his ideas have not caught on in the research on the grammar of PROPS (but cf. Handschuh 2022: 31). In general linguistics, however, the topic has raised some interest. The contributions in Sonnenhauser and Noel Aziz Hanna (2013) are particularly telling because PROPS are mentioned as prototypical hosts of vocatives in most of the articles. In their majority, these studies focus on individual languages. It is true therefore that a properly typological study still seems to be missing. Daniel and Spencer (2009) review the formal properties of vocatives in a variety of languages without specifying for each language which subclasses of nouns boast the vocative. There is evidence from several languages of different affiliation that vocatives are restricted to ANTHS and thus distinguish them from COMMS and TOPOS no matter what coding strategy is employed and independent of the exact status of the vocative.[36]

For formal and theoretical linguists, one type of vocatives is especially interesting because of the strategies employed in its formation, viz. vocative formation via truncation. Blake (1994: 9) mentions the Austronesian language Yapese whose ANTHS have "special forms used for address". More precisely, the formation of the vocative in Yapese requires the reduction of polysyllabic ANTHS to their initial syllable (with or without additional segmental manipulation) (Jensen 1977: 114–116). No other nouny word-class realises the vocative. It is thus a distinctive category of ANTHS. Both the truncation or subtraction strategy and the limitation of the voc-

---

36 The vexing question of old "is it a case, yes or no?" also forms a recurrent theme in Sonnenhauser and Noel Aziz Hanna (2013).

ative to ANTHs are reported repeatedly. Similar patterns are attested throughout the Italian dialect continuum (Thornton 1996). According to Daniel and Spencer (2009: 629), the isolate Nivkh has vocatives for ANTH(-N)s which are formed by dropping the final consonant and vowel lengthening. Alaskan Yup'ik employs truncation exclusively for vocatives of ANTHs (McCarthy and Prince 1986). However, for the purpose of this study, the morphological strategy is of no avail. What counts the most is the fact that there is a functional domain which belongs exclusively to ANTHs.

We find other kinds of vocative with ANTHs which involve different marking strategies. Adelaar (2017: 58) mentions Southern Peruvian Quechua where the suffix -*y* (plus stress shift to the last syllable) marks the vocative on kinship terms and ANTHs referring to "persons of respect." In Lao, the vocative particle *qeej4* follows immediately after the ANTH(-N) of the addressee (Enfield 2017: 193). Aikhenvald (2010: 111) identifies Tariana and Tucano as a further two languages which allow vocatives only with "personal names, kinship terms, and a handful of other nouns with human reference", i.e., vocatives are the privilege of ANTH(-N)s in these languages. In the Dravidian language Koṇḍa

> [p]ersonal names, descriptive titles, kinship terms may be optionally preceded by certain bound or free forms which can be called pre-vocatives [. . .]. Some of these signal the gender of the addressee. (Krishnamurti and Benham 2020: 318)

This is again a case of only ANTH(-N) having access to a dedicated construction for the vocative. There is no need to prolong this list of references any further because it should be clear by now that when searching for properties which characterise ANTHs as different from COMMS vocatives cannot be skipped. Since in many languages, the vocative is also possible for COMMS, an interesting research question arises, namely whether the tetrachoric table (= Table 6) captures the cross-linguistic variation adequately. TOPOS are excluded from Table 6 because of the scarcity of information about vocatives on TOPOS.

Table 6: (Im)possible distribution of vocatives over ANTHs and COMMS.

vocative		ANTH	
		yes	no
COMM	no	yes	yes
	yes	yes	no

The cell which is highlighted in grey represents the impossible combination. We do not expect to find a language which has vocatives for COMMs but none for ANTHs. Thus, an implication can be put forward according to which VOCATIVE$_{COMM}$ ⊃ VOCATIVE$_{ANTH}$. The unmarked status of ANTH-vocatives has a simple functional explanation since vocatives primarily serve to address someone or catch someone's attention. Thus, the vocative is prototypically directed towards a human interactor – and this interactor normally is invested with a distinct ANTH. In Section 3, further cases of distinctive vocatives of PROPs are mentioned.

Parallel to the research on ANTHs in typological perspective conducted in Regensburg, our research-team at the University of Bremen has carried out a number of studies in preparation for our large-scale project *Morphosyntaktische Typologie der Toponyme / Morphosyntactic typology of toponyms (TYPTOP)* to which this investigation also belongs. Three publications must be mentioned specifically:

- Stolz et al. (2014) is a book-length study on zero-marking of spatial relations which originally was not intended as a contribution to the research programme dedicated to the grammar of PROPs. However, the results showed that TOPOs are by far more prone to zero-marking than COMMs (ANTHs were not systematically accounted for) so that another tetrachoric table (= Table 7) makes sense which reflects the implication SPATIAL ZERO-MARKING$_{COMM}$ ⊃ SPATIAL ZERO-MARKING$_{TOPO}$.

Table 7: (Im)possible distribution of spatial zero-marking over TOPOs and COMMs.

spatial zero-marking		TOPO	
		yes	no
COMM	no	yes	yes
	yes	yes	no

- Stolz et al. (2017b) is the first attempt to sketch STG in typological perspective by way of discussing further examples of spatial zero-marking alongside a selection of further grammatical phenomena (proprial articles, genitives, allomorphy of prepositions, blockage of argument status) for which the behaviour of TOPOs (and occasionally that of ANTHs, too) deviates from that of COMMs.
- Stolz et al. (2018) is the direct follow-up to the previous publication. It is shown that spatial zero-marking is cross-linguistically very common but by no means the only peculiarity of TOPOs which sets this class apart from other nouny word-classes. There is also evidence from inflectional morphology.

Section 5 will amply show that inflectional morphology and spatial relations define a functional space in which TOPOs very often display a behaviour that is not (fully) in line with that of ANTHs and/or COMMs. We are now in a position to seriously start with the cross-linguistic investigation of STG.

2.5 Expectations and stipulations

Sections 2.1–2.4.2 have given the reader a foretaste of what is coming. Sections 3–5 carry a heavy load of empirical evidence of SOG. This is meant to hammer the message home that typology will gain new general insights into the structural diversity of languages by way of systematically taking account of what happens in the domain of PROPs. At the same time, an often neglected part of human language may finally be paid the attention it deserves.

However, the empirical part of this study requires several clarifications prior to looking at the data as such. First of all, Table 8 reveals that cross-linguistically we expect to find evidence of languages which (ideally) realise one of five potential types, four of which are directly connected to the existence of SOG.

Table 8: Five possible types.

type	patterns	special grammar(s)
I	COMM = ANTH = TOPO	no
II	COMM ≠ (ANTH = TOPO)	SOG
III	(COMM = ANTH) ≠ TOPO	STG
IV	(COMM = TOPO) ≠ ANTH	SAG
V	COMM ≠ ANTH ≠ TOPO	STG ≠ SAG

We use SOG in the broad sense as the macro-category, under the umbrella of which three different Special Grammars can be chosen from. There is *SOG* (= Type II) in the narrow sense[37] which applies if ANTHs and TOPOs obey the same set of rules which is not shared by COMMs.[38] Strictly speaking, SOG deserves to be looked into in a separate chapter. In the case of **Maori** [Austronesian, Oceanic], Bauer (1993: 255–256)

---

[37] Narrow SOG applies if ANTHs and TOPOs obey identical sets of rules which differ from those which hold for COMMs. Broad SOG on the other hand, is but the cover term for all kinds of special behaviour of PROPs.

[38] A potential Type II-language is Qaqet [Baining]. Hellwig (2020: 79) states that "[t]he morphological patterns [. . .] are not restricted to personal names, but are attested for all proper nouns: personal names of humans, cultural heroes and spirits, place names and names of ethnic groups and languages." Whether the behaviour is the same for all kinds of PROPs also outside morphology is a question that we cannot answer in this study (but cf. Section 3).

takes the distinctive and partly parallel morphosyntactic behaviour of ANTHs and TOPOs to mean that they

> do not belong to the class noun. They differ from nouns in a number of ways, but whether they constitute separate word classes or whether they are sub-classes of the class noun is not a matter easily resolved.

This quote nicely describes what our own project is struggling with. The problems that arise in Maori are by no means unique to this language as this study amply proves. Bauer (1993: 262–263) speaks of personal nouns (= ANTH(-N)s) and local nouns (= TOPO(-N)s). As discussed in Stolz and Levkovych (2022: 253–255), the structural evidence of a separation of PROPs from COMMs is compelling because the latter are always accompanied by the article *te* whereas PROPs require *a* when functioning as subjects. If it were only for this shared property of ANTHs and TOPOs, one could assume that narrow SOG applies. However, ANTHs and TOPOs behave differently from each other (and from COMMs) on other parameters so that it makes more sense to speak of SAG and STG which partly overlap in Maori.

We acknowledge that narrow SOG as such is an interesting subject matter for the future. In this study, however, our focus is on SAG and STG, respectively. To save space, we will refer to narrow SOG only on the side when we discuss instances of STG and SAG. In Type III, STG applies whereas ANTH and COMM form one class. This is different in Type IV where there is SAG but TOPOs and COMMs fall under one and the same rubric. Finally, Type V is the scenario which covers the co-existence of STG and SAG in a given language. This typology is of course far too simplistic and thus calls for further explanations.

For this study, we pass tacitly over Type I, i.e., languages in which all nouny word-classes behave morphologically and morphosyntactically the same. Given the defining traits of PROPs (especially monoreferentiality), it is likely that at least some distributional differences between PROPs and COMMs can be identified no matter which language we look at. Thus, the existence of Type I becomes doubtful.[39] Whether Type I exists crucially depends upon the choice of phenomena which distinguish the different subclasses of nouns. These phenomena range from singularities on the microlevel via those which affect larger numbers of items regularly on the mesolevel to clear-cut and exceptionless behaviour of major classes on the macrolevel. Since this study is explorative and aims at collecting a convincing amount of cross-linguistic data in support of SOG, we not only

---

[39] For Mauwake, a member of the Trans-New Guinea language family, Berghäll (2015: 60–62) emphasises that the structural differences between PROPs and COMMs are minimal, if at all. She rebuts previous accounts in which PROPs and COMMs have been treated as two separate word-classes. PROPs can be modified by adjectives and take determiners even if the reference is unique.

accept all kinds of meso- and macrolevel evidence (except absolute microlevel idiosyncrasies) as valid but we also relegate the discussion of Type I to a follow-up study because Type I-languages (if they exist) would not contribute anything to the data collection.

The deliberately all-embracing approach to the data implies that the abbreviations used in Table 8 have to be understood as follows. The rules which single out PROPs from all other nouns do not have to be mandatory. Often enough they are only optional. Yet, they constitute an option that COMMs usually do not have and this makes them distinctive in the sense of SOG. Moreover, a label like ANTH in the type-defining formulas should not be mistaken for the representative of the entire class of ANTHs. It is by no means exceptional that only a more or less extended subclass of the labelled PROP-class is affected by the SOG rules. Accordingly, ANTH means that there is at least a subset of ANTHs with more than two members to which SOG rules can be applied. That n ≥ 3 is a crucial quantity is only a stipulation of ours to exclude individual idiosyncrasies from overly diversifying the picture. Not only is it possible that the SOG behaviour is restricted to a subset of a subclass of PROPs, but it is also frequently the case that the SOG features are licit only in certain contexts, be they morphosyntactic, pragmatic, or stylistic. To invoke Types II–V, it suffices that there is robust evidence of the recurrence of SOG phenomena in a particular segment of a given language's grammatical system. This also means that if we claim that a given language belongs, for instance, to Type III we knowingly simplify the complex situation. The complexity of the situation is rendered even more complex by the possibility that in one and the same language, ANTHs may side with TOPOs on parameter A whereas they go along with COMMs on parameter B while TOPOs are like COMMs and/or ANTHs on parameter C, etc.

To get going with the project on the empirical side, we postpone the solution of the above methodological and theory-related problems until a later stage when the database is sufficiently big to allow us to separate the wheat from the chaff, in a manner of speaking. For the time being, we concentrate on hunting and gathering data. To this end, the Working Hypothesis in (27) is put forward to guide us through the upcoming parts of this study.

(27) Working Hypothesis
PROPs attest to SOG especially in those functional domains in which they are prominently involved, i.e.,
(a) ANTHs are prone to differ morphologically and/or morphosyntactically from COMMs (and TOPOs) in verbalised situations where the name-bearing participants fulfil prototypical functions, namely those which require a human participant to take on a specific role such as the possessor in adnominal possession,

(b) TOPOS are prone to differ morphologically and/or morphosyntactically from COMMS (and ANTHS) in verbalised situations where the name-bearing participants fulfil prototypical functions, namely those which require a geo-object to function as Ground for a Figure.

In Section 4, part (27a) of the Working Hypothesis will be important whereas (27b) is crucial for the discussion in Section 5. To show that the functional domains mentioned in the Working Hypothesis do not cover the SOG phenomenology in its entirety, Section 3 hosts a selection of statements on SOG features from an array of descriptive grammars which do not always elaborate on the points of interest.

3 Empirical bits and pieces

This section exclusively contains examples drawn from descriptive grammars. Phenomena which have already been discussed in the literature on the grammar of PROPS are left out. The list of consulted grammars is of course not exhaustive. The data are presented with the intention of raising the interest of fellow-linguists in the study of SOG. In our sources, the examples more often than not are not further elaborated upon so that they are not fit for being presented in the format of dedicated case-studies here. The examples are given in the alphabetical order of the glossonyms.[40] The focus is on the distinctiveness of certain SOG-features, not all of which are inflectional or morphosyntactic. The number of languages has arbitrarily been limited to fifteen. No attempt has been made at creating a balanced sub-sample. The sketches come in different sizes according to the availability of sentential examples in the descriptive grammars.

Abkhaz [Abkhaz-Adyghe] displays two properties which speak in favour of the existence of SOG. According to Hewitt (1989: 44–45), Abkhaz boasts an associative plural encoded by the suffix *-ra:* which is limited to ANTHS as in *Zaə̀ra-ra:* 'Zaira **and family/friends/company**'. In addition, PROPS lack the otherwise obligatory article *a-* (Hewitt 1989: 45). It remains unclear whether ANTHS and TOPOS behave the same. In any case, there is evidence of SAG.

Cantonese [Sino-Tibetan, Sinitic] has a prefix *a-* whose distribution is restricted to ANTHS and kinship terms, the function of which is to signal familiarity (Matthews and Yip 2011: 43). The prefix is not mandatory. Its use with kinship terms seems to be particularly frequent with expressions referring to members of the older generations (Matthews and Yip 2011: 432). The three sentences under (28) show that, depending on definiteness, COMMS are usually preceded by a classifier like the generic *dī* in (28a). In (28b), we see an example of an ANTH hosting the prefix *a-* whereas the TOPOS in (28c) are accompanied neither by classifiers nor by the familiarity prefix *a-*.

---

[40] The focus being on morphological and morphosyntactic issues, we do not systematically account for phonological schema preservation (Nübling 2017a). This does not mean that there is no compelling cross-linguistic evidence of the phenomenon as cases like Belep [Austronesian, Oceanic] where medial voiceless stops in PROPS escape the usual sonorisation which is mandatory with COMMS (McCracken 2021: 115). We strongly recommend that this and related phenomena is studied in-depth separately in cross-linguistic perspective.

(28) Cantonese [Sino-Tibetan, Sinitic]
　(a) COMM (Matthews and Yip 2011: 164)
　　dī　　sailouhjái　jáu-jó　　yahp　heui
　　CL　　child.PL　　run-PFV　in　　go
　　'The children came running in.' [O.T.]
　(b) ANTH (Matthews and Yip 2011: 434)
　　a-Lìhng　　m̀h　hang　　ga　　béi　ngóh
　　ANTH-Ling　NEG　willing　marry　to　　1SG.OBJ
　　'Ling is not willing to marry me.'
　(c) TOPO (Matthews and Yip 2011: 194)
　　Wāngōwàh　móuh　　　Hēunggóng　gam　bīkyàhn
　　Vancouver　　NEG.have　Hong Kong　as　　crowded
　　'Vancouver is not as crowded as Hong Kong.' [O.T.]

There is thus a tripartition of the macro-class nouns into three subclasses with different structural properties. Cantonese attests to the co-presence of SAG and STG according to the formula introduced for Type V.

Chechen [Nakh-Daghestanian, Nakh] shows allomorphy in the ergative. Nichols (1994: 24) lists four co-existing ergative suffixes, namely -*uo*, -*ie*, -*s*, and -*a* with -*s* being "used on personal names and some kin terms." The choice of suffix shows that ANTH-Ns behave differently from other nouns. Chechen thus gives evidence of SAG of Type IV. The mismatch manifests itself in the existence of inflectional classes, one of which being reserved for ANTH-Ns.

The grammarians of **Chukchi** [Chukotko-Kamchatkan] claim that the division of the nouns of this language into two classes is morphologically relevant (Kämpfe and Volodin 1995: 37). There is the class of ANTH-(N)s which is distinct from that of COMMs to which the vast majority of nouns belong. ANTH(-N)s form the 2nd declension with two subclasses: (a) is reserved for ANTHs and (b) contains kinship terms (Kämpfe and Volodin 1995: 85). The 2nd declension differs from the 1st declension insofar as the former has dedicated affixes for the opposition COLLECTIVE ≠ NON-COLLECTIVE which are illicit with COMMs (Kämpfe and Volodin 1995: 82–83). This is a case of **overdifferentiation**. Since the marker of [+collective] is compatible with the case postfix in the ablative, allative, and orientative, some word-forms of ANTH(-N)s count more slots than the COMMs. Since Kämpfe and Volodin (1995: 82) assume that the overt coding of case and number is blocked for word-forms of Chukchi nouns, they rebut the idea that the collective is a number category. No matter whether the affixal morphology of Chukchi might better be analysed differently, one fact remains uncontroversial, namely that Chukchi represents SAG of Type IV.

For **Coastal Marind** [Anim], Olsson (2022: 49) states that

[p]roper names share most of their distributional possibilities with standard nouns: for example, they can be determined by demonstratives (*Maria upe* 'that Maria') and occur in possessive constructions with *en* 'POSS' (*Wodim en Maria* 'Wodim's [daughter] Maria'). The main difference with lexical nouns is that proper names may occur with the Associative Plural *ke* or *keti* [...], a property they share with kinship terms.

TOPOS are not mentioned. On the basis of the above quote we hypothesise that Coastal Marind gives evidence of SAG of Type IV.

Dogon [Dogon] employs the clitic *ŋ* as marker of direct and indirect objects as well as marker of predicate nouns. The latter function seems to be possible with all subclasses of nouns whereas object-marking is strictly limited (apart from personal pronouns) to ANTH(-N)s (Plungian 1995: 12–15) as shown in (29).

(29)   Dogon [Dogon]
   (a)   COMM as object (Plungian 1995: 11)
       gamma   gɛ   ay   Ø   aw-e-Ø
       cat   DEF   mouse   Ø   catch-AOR-3SG
       'The cat has caught a mouse.'
   (b)   ANTH as object (Plungian 1995: 12)
       Sana   Kanda   ŋ   bo-e-Ø
       Sana   Kanda   OBJ   call-AOR-3SG
       'Sana has called Kanda.'

In (29a), the object function of the COMM *ay* 'mouse' is not marked overtly whereas the presence of the object marker *ŋ* is obligatory in (29b) because the object is an ANTH. The use of the object marker is not licit with COMMs. To describe the morphosyntactic behaviour of COMMs and ANTHs in the domain of object-marking, two (sub-)rules have to be postulated – and this necessity speaks in favour of considering this a case of SAG according to Type IV (no information as to the behaviour of TOPOS is given in the source).

Treis (2008: 108–113) gives a detailed account of the grammar of PROPs in **Kambaata** [Afro-Asiatic, Cushitic]. She observes that PROPs "differ from the vast majority of common nouns with respect to case and gender marking as well as with respect to copula use." Inflectionally, ANTH(-N)s differ from COMMs insofar as they display no secondary gender markers in the nominative and accusative. Where COMMs like NOM *lókk-a-t* {ear}-{NOM}-**{F}** 'ear' / ACC *lokk-á-ta* {ear}-{ACC}-**{F.ACC}** host an overt gender marker for feminine, female ANTHs do not, e.g., NOM *Bóoy-a* / ACC *Booy-á* (Treis 2008: 103–104). In (30), it is shown that in copula constructions ANTHS as complements require the use of Copula 3 (30a) whereas COMMs take Copula 2 (30b).

(30) Kambaata [Afro-Asiatic, Cushitic]
   (a) ANTH (Treis 2008: 422)
   Boq-é   hiz-óo   Báaf-aa-t
   Boqe-M.GEN sibling-M.NOM   Baafa-PRED-**COP3**
   'Boqe's brother **is** Baafa.' [O.T.]
   (b) COMM (Treis 2008: 411)
   *Ku   bún-u   gummúut-a-**a***
   DEM.PRX   coffee-M.NOM   lukewarm-M.PRED-M.**COP2**
   'This coffee **is** lukewarm.' [O.T.]

In point of fact, Copula 2 is also admissible for a small selection of COMMs so that it is more precise to speak of ANTH(-N)s taking Copula 2. The same Copula 2 is also required by pronouns. Since TOPOs are said to take an intermediate position between COMMs and ANTHs in the domain of morphology (Treis 2008: 112), we assume that the language represents Type V with coexisting SAG and STG.

In **Kera** [Afro-Asiatic, Chadic], TOPOs may host the pluralising prefix *kə- ~ gə-* to form a noun which refers to the inhabitants of a given settlement as in *Mə́tá* → *kə-Mə́tá* {**PL**}-{Mə́tá} '**people from** Mə́tá'. The bare TOPO cannot refer to a single inhabitant of a settlement. In order to express this notion, it is necessary to form a prepositional attribute whose complement is the TOPO as in *hùlùm [bə̀ gùbùsíyá]*_{PP} {man} [{from} {Gùbùsíyá}]_{PP} 'a man from Gùbùsíyá' (Ebert 1979: 151). More generally, pluralisation in Kera is restricted to human nouns, economically/culturally important animals, and certain artefacts (Ebert 1979: 150).

Postpositions in **Kharia** [Austroasiatic, Mundaic] "generally take the genitive with proforms and proper names and the direct case with other semantic bases" (Peterson 2011: 150). This rule also holds for complex postpositions although in this case, the genitive seems possible also with some COMMs. Thus, ANTHs and COMMs differ as to which case is required of them when they function as complement of a given postposition. Interestingly, ANTHs opt for the same solution as pronouns.

Koasati [Muskogean] is a language in which ANTH(-N)s stand out from the other nouny word-classes because of two properties. On the one hand, they form the Noun Class V whose separation from the other four noun classes rests on the existence of a formally distinct vocative (Kimball 1991: 390). This PROP-vocative is formed by way of dropping the final vowel of the PROP or by truncating the derivational suffix *-ka* as, e.g., *Simantá:li* (male PROP) → VOC *Simantá:l*. COMMs employ the so-called autonomous case (identical with the citation form) as vocative (Kimball 1991: 402–403) as, e.g., AUTONOMOUS CASE = VOC *alikcí* 'doctor ~ doctor!'. ANTH(-N)s are responsible for the existence of a fifth inflectional class in Koasati noun morphology and in this way, they add to the noncanonicity of the system. The second property of interest is the function of the suffix *-o:to* 'the deceased; the long ago' which Kimball (1991: 409) subsumes under the article suffixes of Koasati.⁴¹ It can be used on PROPS as well as on COMMS but with different meanings. According to Kimball (1991: 409)

> [i]t has two uses: suffixed to nouns referring to specific persons, it indicates that the person being spoken of is dead [...]; suffixed to other nouns, it locates them in the past and indicates that the action they participate in is true only for that time in the past [...].

The different readings of the suffix can be gathered from the examples in (31).

(31) Koasati [Muskogean]
    (a) PROP (Kimball 1991: 409)
        Soy-ó:to-k   *ónka-k*   *akakanahilká*   *ká:ha-V́hco-toho-n*
        Soya-**deceased**-SBJ   quoth-SS   looking_glass   say-HABIT-RLS-DS
        'Soya, **who is now deceased**, used to say "looking-glass".' [O.T.]
    (b) COMM (Kimball 1991: 410)
        wacin-ó:to-n   *ca-sobáy-ko-k*
        English-**long_past**-OBJ   1SG.STAT-know-3.NEG-SS
        '[...] **at that time**, I was ignorant of English [...].' [O.T.]

TOPOS are not discussed in the descriptive grammar. We assume that Koasati attests to SAG by way of realising the pattern of Type IV.

Mian [Nuclear Trans New Guinea, Ok] is another language which gives evidence of SAG. Together with kinship terms, ANTHS form the class of ANTH(-N)S which differs from COMMS insofar as only ANTH(-N)S can take the vocative clitic =o (as in *Kasening=o* 'Kasening!') and the suffix -*wal* which expresses the Associative Plural with ANTHS (as in *Kasening-wal* 'Kasening and his people/family'). Attached to kinship terms, -*wal* has two possible readings, namely as marker of the Associative Plural (*awók-mal* 'the mother and her people/family') or of the genuine additive plural (*awók-wal* 'the mothers') (Fedden 2011: 95). Note that COMMS are pluralised not by way of suffixation but by way of using dedicated articles (Fedden 2011: 86). In the absence of specific information about the morphosyntactic behaviour of TOPOS, we conclude that Mian, too, is a language with SAG of Type IV.

In Taylor's (1985: 82, 84) grammar of **Nkore-Kiga** [Atlantic-Congo, Bantu] the vocative is mentioned as a category in which a difference between PROPS and COMMS comes to the fore. To form the vocative, COMMS are subject to aphaeresis, i.e., they lose their initial vowel. This rule does not affect PROPS because their first segment is always a consonant. The initial vowel of COMMS is an antefix attached to

---

41 Becker (2021: 41–42) shows that the Koasati suffix does not fulfil her criteria for articlehood.

the left of a regular class prefix. Under certain syntactic conditions this vowel can be dropped (Taylor 1985: 88–89).

We also witness the presence of SOG in **Qaqet** [Baining]. Hellwig (2020: 80–83) argues that all PROPS share formal properties especially in the domains of anaphora and agreement. In the case of COMMS, agreement is grammatical, i.e., according to formal criteria of the controller. With PROPS, however, agreement is semantic – ANTHS which have a female referent trigger feminine agreement whereas ANTHS with male referents require masculine agreement. This is remarkable in those cases where the ANTH is invested with a noun class suffix which normally determines the agreement morphology on the targets as shown in (32).

(32) Qaqet [Baining] (Hellwig 2020: 81)
 *ma=[genainymet-**ki**-ni]$_{ANTH}$*   **ke**=*rarles*
 prop=Genainymet-SG.**F**-SG.DIM   3SG.**M**.SBJ.NPST=start.NCONT
 *i=**ke**=ksik*
 SIM=3SG.**M**.SBJ.NPST=climb.CONT
 'Genainymetkini started to climb [. . .].' [O.T.]

The ANTH *Genainymetkini* involves the feminine singular suffix *-ki*. The overt marking for feminine gender, however, is overruled by the referent's sex so that the person marker *ke=* of the 3rd person singular masculine is used on both verb forms. Similarly, "[p]lace names belong to the same class as their superordinate category, irrespective of their overt marking" (Hellwig 2020: 81). This means that PROPS follow their own agreement rules which do not correspond to those which apply in the case of COMMS. What is more, in contrast to COMMS, PROPS are excluded from combinations with the noun marker *a* and the article *ama*. PROPS take *ma* (which can optionally be dropped in the vocative) (Hellwig 2020: 81). In Hellwig's terminology, *ma* is the article of inherently identifiable referents which "occurs especially with proper nouns" (Hellwig 2020: 132) and it is obligatory with TOPOS (Hellwig 2020: 133). We therefore classify *ma* as PROP-article (with the proviso that article is probably not the best choice of term). The usage of the articles is illustrated in (33).

(33) Qaqet [Baining]
 (a) COMM (Hellwig 2020: 128)
  ma=siqi   *de*   ***ama**=qaqera-ki*   ***ama**=tlu-ki*
  prop=Siqi   CNJ   **ART**=person-SG.F   **ART**=good-SG.F
  'Siqi, she is a good woman.' [O.T.]

(b) ANTH (Hellwig 2020: 132)
   murl   ***ma**=sirini*   *ka=tit*
   distantly   **PROP**=Sirini   3SG.M.SBJ=go.CONT
   'In the past, Sirini was going.' [O.T.]
(c) TOPO (Hellwig 2020: 133)
   kurli-ut=iara   *pet*   ***ma**=rlaunspna*
   leave-1PL=PRX   on   **PROP**=Raunsepna
   'We stay here in Raunsepna.' [O.T.]

Since the classes of ANTH and TOPO seem to share most of the traits which distinguish them from COMMs, it is possible that Qaqet is a language of Type II, i.e., an instance of SOG as already mentioned in footnote 38.

Tsukida (2011: 301) states that **Seediq** [Austronesian, Atayalic] has a small system of cases. The bulk of the nouns distinguish the unmarked direct case (which also serves as citation form) from the nominative which is preceded by the particle *ka*. This is the situation in the domain of COMMs. However, when we look beyond COMMs we notice that there is another class of nouns whose members take the additional suffixed oblique marker *-an*. This property is reserved for ANTH(-N)s such as *Rubiq* → OBL *Rubiq-an* (including the word *se'diq* 'people'). Note that the oblique in *-an* is also attested with pronouns. ANTH(-N)s additionally boast a vocative (unattested with COMMs) which is formed by reducing the phonological chain of the ANTH(-N) to its final syllable as in *Masaw* → VOC *Saw!* Whether the prefixation of *ne-* to nouns to form possessive nouns like *Rubiq* → *ne-Rubiq* 'what belongs to Rubiq' is a privilege of ANTH(-N)s or is possible for animate nouns in general or for a subset thereof remains unclear.

The vocative is crucial for SOG also in the case of **Somali** [Afro-Asiatic, Cushitic]. Saeed (1999: 64) makes a distinction between "vocatives" and "noun vocatives", the former occurring with ANTH(-N)s and the latter being licit with both ANTH(-N)s and COMMs. Those vocatives which are restricted to ANTH(-N)s are formed by accent shift whereas the so-called noun vocatives are suffixal. In Table 9, it is shown that the suffixes differ between ANTH(-N)s and COMMs (Saeed 1999: 65).

Table 9: Vocative suffixes in Somali.

gender/number	ANTH	COMM
F.SG	-èey ~ -àay ~ -òoy	-yahay
M.SG / PL	-òw	-yohow

The accentual shift in vocative formation for ANTHs can be illustrated by the male ANTH *Warsáme* with high tone on the penultimate in the absolute case as opposed to *Wársame* with high tone on the antepenultimate in the vocative (Saeed 1999: 66).

To sum up the sketches of SOG in the foregoing paragraphs, we emphasise that the superficial diversity notwithstanding, there are also a number of commonalities which connect several of the instances to each other. On the one hand, we have seen that SOG phenomena can be derivational (Cantonese, Chukchi(?), Kera), inflectional (Abkhaz, Chechen, Chukchi(?), Kambaata, Koasati, Mian, Nkore-Kiga, Seediq, Somali), or morphosyntactic (Dogon, Kharia, Qaqet). On the other hand, different grammatical categories are affected such as case (Chechen, Dogon, Koasati, Seediq, Somali), class (Cantonese) and gender (Kambaata, Qaqet), number (Abkhaz, Chukchi(?), Coastal Marind, Mian), and predication (Kambaata). Some of the above languages attest to several SOG-phenomena. In many cases, ANTHs have been shown to behave differently from COMMs to give rise to SAG. Accordingly, Type IV is attested particularly often. STG in the form of Type III is attested only in Kera whereas Kambaata seems to give evidence of Type V. In all cases reported above, SAG and/or STG diversify the grammatical system of the sample languages structurally in the sense that special rules have to be formulated to capture their behaviour. Those rules which are valid for COMMs cannot simply be applied to PROPs. As a consequence, mismatches arise since the existence of different inflectional classes or paradigms with unequal numbers of cells have to be assumed. Different coding strategies for functionally identical categories are at work, too. Thus, the integration of PROPs into the research programme of Canonical Morphology promises a particularly rich harvest for those who take an interest in noncanonicity. Further cases of this kind will be discussed in more detail in Sections 4–5.

4 Leitmotifs of *Special Anthroponymic Grammar*: Possession

This section includes several case studies of varying length which have a common theme. It is shown that ANTHs are especially prone to deviate from those patterns which are valid for COMMs when it comes to expressing adnominal possession. The recurrence of the deviation in exactly this functional domain can be empirically substantiated across many languages. When Davies (1989: 104) states that, in **Kobon** [Nuclear Tans New Guinea, Madang], PROPs can be distinguished from COMMs and kinship terms because PROPs cannot function as possessee in adnominal possession this criterion can hardly surprise us since taking a genitive attribute is a case of modification and modification with PROPs is generally believed to be subject to constraints (without being absolutely impossible). However, PROPs may function as possessors, i.e., as genitive attributes in adnominal possession – termed anchors by Koptjevskaja-Tamm (2002). And it is here where interesting facts come to the fore. Figure 2 identifies the scenarios which are of interest to us in this section.

$$\text{POSS} \rightarrow \begin{cases} X \quad / \text{POSSESSOR} = \begin{Bmatrix} \text{ANTH} \\ \text{PROP} \end{Bmatrix} \\ Y \quad / \text{POSSESSOR} = \text{COMM} \end{cases}$$

Figure 2: Schematic distinction of possessors according to word-class.

In a given language, there are two (or more) constructions available for adnominal possession of the noun-noun kind. Construction X is chosen if the possessor is an ANTH (or PROP) whereas construction Y is activated if the possessor belongs to the class of COMMs. Clearly, the rule is overly schematic so that many aspects are glossed over. The simplification notwithstanding, the split rule in Figure 2 captures the most important basics of the cases we discuss in the subsequent sections. In Section 4.1, we present data from a variety of languages with different genetic affiliation, geographical location, and structural properties. These are languages for which only relatively short sketches along the lines of those presented in Section 3 are provided. Sections 4.2–4.4 are dedicated to a few Indo-European languages of Europe to whose SAG properties Stolz et al. (2017a) already referred briefly. We take up the issues raised in the previous publication by way of scrutinising the data more thoroughly. Section 4.5 recapitulates our findings, determines what they mean for the linguistic theory of possession, and explains how they connect to the topic of Section 5.

4.1 Familiarising ourselves with SAG in the domain of possession

What is not a case of SAG? Superficially, **Manx** [Indo-European, Celtic] seems to fulfil the necessary criteria to be classified as a language which attests to SAG in the possessive domain. However, on closer inspection, it turns out that this interpretation is wrong. We mention this case at this point in order to avoid the danger of lumping together like with unlike. Manx is an extinct language in the process of revitalisation. The descriptions of Manx do not always agree on all points of grammar. Both Draskau (2008: 28) and Broderick (2010: 313) assume that PROPs undergo lenition of the initial consonant when they function as possessor in adnominal possession as in *Juan* /**d'**uən/ 'John' → *thie Yuan* /tai juən/ 'John's house' and *Moirrey* 'Mary' → *mac Voirrey* 'Mary's son'. The lenition is the only exponent of the genitival relation. The two sources disagree as to the possibility of lenition serving the same purpose outside the domain of ANTH-possessors. Draskau (2008: 28) assumes that definite masculine COMMs in the singular are also lenited in possessor function whereas Broderick (2010: 313) mentions neither definiteness nor singular as criteria for the lenition of masculine COMMs in this function. Whatever the solution to this problem might be, we can cancel Manx from the list of SAG-candidates because ANTHs share the above property with a larger segment of the COMM class which cannot be subsumed under ANTH-N. Figure 3 is meant to illustrate the overlap of COMMs and ANTHs in Manx.

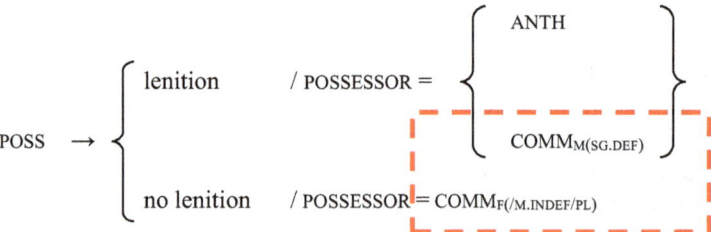

Figure 3: Schematic distinction of possessors according to word-class in Manx.

The box (delimited by the red interrupted line) indicates that the class of COMMs is divided in two with one major segment siding with ANTHs as to being lenited under possession. ANTHs are thus not sufficiently distinct from COMMs to deserve the label SAG.

Now that we know that cases of the Manx kind do not constitute SAG we can look at bona fide instances of SAG in the context of possession.

In **Amele** [Madang, Gum], PROPs do not allow for being modified and thus fail to co-occur with attributive nouns, adjectives; they are also excluded from combinations with indefinite articles, quantifiers, and markers of emphasis. All these contexts are reserved for COMMs. In contrast, "[a] proper noun can also occur as the object of the postposition in the possessive postpositional phrase whereas a common noun cannot" (Roberts 1987: 152). In this domain, PROPs behave like pronouns and kinship terms. Possessor and alienable possessee are linked to each other by the postposition *na* 'of' as in (34a). This construction frame being blocked for COMMs, adnominal possession is expressed by juxtaposition of possessor and possessee as in (34b) (Roberts 1987: 139).

(34)   Amele [Nuclear Trans New Guinea, Madang] (Roberts 1987: 171)
    (a)   possessor = ANTH
        Banang **na**   *jo*   *mane-i-a*
        Banang GEN house burn-3SG-HODPAST
        'Banang's house burned down.' [O.T.]
    (b)   possessor = COMM
        [dana   caub]$_{POSSESSOR}$ Ø   *caja$_{PSM}$*   *ho-na*
        man   white   Ø   woman   come-3SG.PRS
        'The white man's wife is coming.' [O.T.]

The two different constructions can be captured by a split rule as shown in Figure 4 (pronouns are skipped).

$$POSS \rightarrow \begin{cases} na & / \text{ POSSESSOR} = \text{PROP} \\ \\ \emptyset & / \text{ POSSESSOR} = \text{COMM} \end{cases}$$

Figure 4: Split rule in Amele.

The opposition between PROPs which require overt marking of the possessive relation and COMMs which allow for zero-marking is especially telling for the topic under inspection. SOG does not always manifest itself in the absence of phonologically realised markers. To the contrary, SOG – and accordingly SAG and STG, too –

frequently requires the use of distinct dedicated markers. The phenomenology of SOG in general is not defined only ex negativo.

In **Kryts** [Nakh-Daghestanian, Lezgic] the inflectional genitive of PROPS is different from that of COMMS. Authier (2009: 32–33) shows that nouns with a high degree of referentiality (i.e., PROPs including kin terms) have zero-marking in the genitive (as, e.g., *Majlis-Ø k'ul* 'the house **of** the Maijlises') whereas COMMS take their pick from the allomorphs *-a, -d, -n, -l, -r,* and *-rd* or vowel changes (as, e.g., *lem-ird yak* 'the meat **of** a donkey' (Authier 2009: 37)). The class of PROP-NS consists not only of ANTHS, TOPOS, and kinship terms but also host nine monosyllabic COMMS which refer to geo-objects. Moreover, many loanwords (all of them COMMS) are admitted to this class. It is therefore possible that the latter component will become dominant in the future so that the class might lose its SOG-character. Note that ANTHS and TOPOS behave in identical ways since both lack a phonologically realised genitive exponent (Authier 2009: 47). This can be counted as evidence of SOG (= Type II).

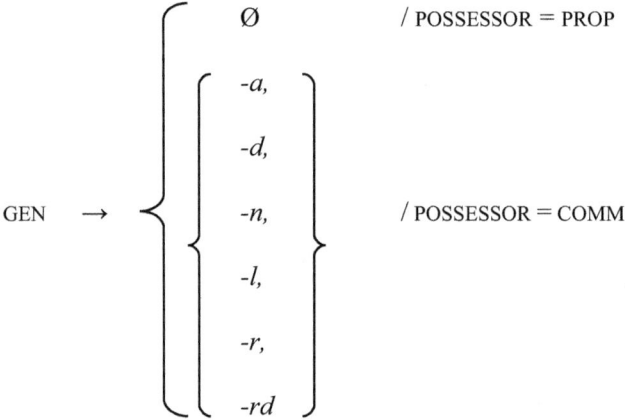

Figure 5: Split rule in Kryts.

Since the paradigms of PROPS are characterised by absolutive-genitive **syncretism** (both cases are zero-marked) they differ markedly from COMMS whose paradigms always host formally distinct absolutives and genitives. This means, that PROPS are responsible for the existence of different inflectional classes in this language.

In **Malagasy** [Austronesian, Basap-Greater Barito] COMMS and ANTHS do not only differ as to the choice of article – COMM *ny* ~ *ilày* ≠ ANTH *i* – but also with regards to the exponent of the genitive (Rasoloson and Rubino 2011: 466). The genitive markers come in two sets, namely
(a) the proclitics *an'ny* ~ *an'ilay* for COMMS and *an'i* for ANTHS and
(b) the enclitics *=n'ny* for COMMS and *=n'i* ~ *=n* for ANTHS.

In (35), the use of the proclitics is illustrated. *an'ny* and *an'i* are phonologically distinct because the former has a medial geminate nasal /nn/ where the latter makes do with the singleton /n/.

(35) Malagasy [Austronesian, Basap-Greater Barito]
 (a) COMM (Rasoloson and Rubino 2011: 466)[42]

an'ny	*mpampiànatr'*	*i*	*Sèndra*	*'ty*	*pôkètra*	*'ty*
GEN.COMM	teacher.GEN.ANTH	ANTH	Sendra	PRX.VIS	bag	PRX.VIS

'This bag belongs to Sendra's teacher.' [O.T.]
 (b) ANTH (Rasoloson and Rubino 2011: 467)

an'i	*Rina*	*'zào*	*e!*
GEN.ANTH	Rina	now	INTJ

'It is Rina's (turn) now!' [O.T.]

The combination of two possessive relations in (35a) already tells us that the story of the Malagasy genitive is not that simple. Rasoloson and Rubino (2011: 468) report that "a number of different scenarios have to be distinguished" if the possessor is an ANTH or a definite NP. First of all, the default rule requires that the enclitics =*n'ny* and =*n'i* (which are phonologically indistinguishable /ni/) are attached to the possessed noun. At this stage, the distinction COMM ≠ ANTH is merely orthographical as in

(a) *ny bòki=**n'ny** mpampiànatra* {COMM} {book}={**GEN.COMM**} {teacher} 'the teacher's book(s)' vs
(b) *ny kiràro=**n'i** Fàly* {COMM} {shoe}={**GEN.ANTH**} {Faly} 'Faly's shoes.'

However, in contrast to cases which involve a COMM as possessor, ANTH-possessors are responsible for the application of additional rules. If the ANTH has an initial rhotic /r/, the genitive enclitic on the preceding possessee is represented by the reduced allomorph -*n* which, in turn, triggers the prosthesis of /d/ on the possessor nouns as in *sàry* 'photograph' + *Rasòlo* → *sari-**n** **d**Rasòlo* 'Rasòlo's photograph'. The allomorph -*n* is also required if the genitival attribute is an indefinite COMM as

---

42 On the basis of the further light shed on the Malagasy genitive by Rasoloson and Rubino (2011: 468), we have reanalysed this example as follows: in the original, the ANTH-article *i* which introduces the ANTH *Sèndra* was presented as resulting from the raising of the final low vowel of the COMM *mpampiàntra* 'teacher'. The replacement of final /a/ with /i/ is possible only if the genitival attribute is a definite COMM. Since *Sèndra* is an ANTH, the example had to be reinterpreted as reflecting the rule which requires apopcope of word-final /a/ in weak syllables of possessee nouns and at the same time the use of the ANTH-article *i* on the possessor (cf. below).

in *tràno* 'house' + *andrìana* 'nobleman' → *tràno-n' andrìana* 'a nobleman's house'. Furthermore,

> if the genitive argument is a personal name and the head ends in a weak syllable (*ka, tra,* or *na*) the vowel *a* of the weak syllable is dropped [...] and no special genitive marker is used: *kàvina* 'earring' > *kàvin' i Rìna* 'Rìna's earring(s)'. [...] [T]his rule also applies to personal names beginning with /r/, which then are preceded by the personal article *i* [original boldface and italics] (Rasoloson and Rubino 2011: 468).

The apocope of the final /a/ is blocked if the possessor is not an ANTH but a definite COMM. In this case, /a/ is replaced with /i/ as in
- *pèratra* 'ring' → *pèratr-y ny rahavàvi=ko* {ring}-**{GEN.COMM}** {COMM} {sister}={POSS.1SG} 'my sister's ring'.[43]

Two further properties let ANTHs stand out, namely
(a) the existence of the ANTH-prefixes *I-, Ra-, Ilai,* and *Ikàla-* which block the use of the ANTH-article *i* and
(b) the associative-plural article *ry* which is additionally used when directly addressing individuals (as a kind of vocative article) (Rasoloson and Rubino 2011: 466).

Figure 6 summarises the intricate rule split for the genitive constructions.

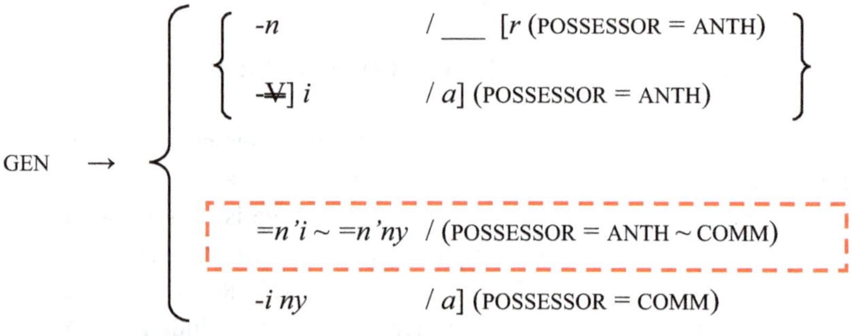

$$\text{GEN} \rightarrow \begin{cases} \begin{cases} -n & / \underline{\quad} [r \text{ (POSSESSOR = ANTH)} \\ -\cancel{V}] \, i & / \, a] \text{ (POSSESSOR = ANTH)} \end{cases} \\ =n'i \sim =n'ny \;\; / \text{ (POSSESSOR = ANTH} \sim \text{COMM)} \\ -i \, ny \quad\quad\quad / \, a] \text{ (POSSESSOR = COMM)} \end{cases}$$

Figure 6: Split rule in Malagasy.

The box highlighted in red marks the zone of overlap shared by ANTHs and COMMs. In contrast to the situation in Manx (Figure 3), the overlapping segment is counterbalanced by several other SAG-subrules.

---

[43] In the original, the possessive clitic *=ko* of the 1st person singular is erroneously glossed as 3rd person singular.

In the Japanese dialects of **Tsuruoka** and **Ei** [Japonic], COMMs take the genitive marker -*N* after vowels and -*no* elsewhere as in *hana-N iro* 'colour of a flower' and *un-no iro* 'colour of an ocean'. However, Matsumori and Onishi (2017: 335) state that "first- and second-person pronouns, individual names, kinship terms, titles referring to a specific individual" take the genitive marker -*ŋa* as in *joʀko-ŋa hon* 'Yōko's book'. Note that in the Ei dialect -*ŋa* also encodes the nominative (Matsumori and Onishi 2017: 334). There is thus **syncretism** of nominative and genitive in this variety of Japanese. Since no information about TOPOs is given in the source we have consulted we conclude that in these regional varieties of Japanese, SAG applies – probably reflecting Type IV. Figure 7 presents the split rule we assume to hold for this case.

$$\text{GEN} \rightarrow \begin{cases} \textit{ŋa} & / \text{ POSSESSOR = ANTH-N} \\ \textit{no} & / \text{ POSSESSOR = COMM} \end{cases}$$

Figure 7: Split rule in Japanese dialects.

The foregoing paragraphs have shown that SAG (sometimes within SOG) is not an absolute cross-linguistic rarity. Sections 4.2–4.4 are intended to prove that phenomena pertinent to SAG can be found also on our linguistic doorstep, in a manner of speaking. Several members of the Germanic phylum as well as Romanian give evidence of SAG in the domain of adnominal possession. These cases will be scrutinised more closely to the benefit of two research programmess, viz. the grammar of PROPs and the linguistic possession research.

4.2 Old genitives and new possessives

4.2.1 Varia Germanica

Germanic genitives have recently gained heightened attention as reflected by the collection of articles edited by Ackermann et al. (2018). PROPs are mentioned time and again on many pages throughout the edited volume for instance in Ackermann's (2018: 199–212) inquiry into the diachronic developments in the history of German although PROPs do not exhaust the phenomenology of the processes studied by the author. The German genitive is prominently featured twice in Nübling et al. (2015: 68–71, 84–86). There is already an abundance of voices which have commented upon the differential behaviour of this case with PROPs and COMMs. The main results of these previous studies are summarised by Nübling et al. (2015) as follows (glossing over many details). In the course of the history of German, the erstwhile rich inventory of genitive markers has been reduced drastically for PROPs.⁴⁴ In the past, even PROPs were distributed over different declension classes. In contemporary German, however, the genitive suffix -s is the only option for PROPs irrespective of gender (e.g. *Marias-s Geburtstag* 'Mary's birthday' (Nübling et al. 2015: 70)) in contrast to COMMs which (in correlation to inflectional classes and gender) preserve several options (such as *der Geburtstag des Junge-n* 'the boy's birthday').⁴⁵ On the phrasal level, the genitive is marked only once if a PROP is involved (as in *Peter-s Geburtstag ~ der Geburtstag Peter-s ~ der Geburtstag des Peter* 'Peter's birthday' (Nübling et al. 2015: 68)), i.e., agreement is blocked principally whereas agreement is still possible in constructions involving a COMM (e.g. *der Geburtstag des Schüler-s* 'the pupil's birthday'). This means that in the domain of inflection and agreement, PROPs are progressive whereas COMMs are conservative. In syntax, however, it is exactly the other way around. PROPs in possessor function may occupy the slot to the left of the possessee (which was mandatory also for COMMs in older stages of German) whereas COMMs as possessors are admitted in this position only exceptionally under particular stylistic conditions. The order POSSESSOR > POSSESSEE is still normal with PROP-possessors (e.g. *[Anna-s]$_{POSSESSOR}$ Sprache* 'Anna's language' (Nübling et al. 2015: 85)) but highly marked with COMM-possessors. The inverted order POSSESSEE > POSSESSOR is the default linearisation for COMM-possessors (e.g. *die Sprache [des Einwanderer-s]$_{POSSESSOR}$* 'the immigrant's language'). The possessor preferably follows the possessee also in the case of PROP-possessors if the possessor is phrasal or consists of more than one word (e.g. *der Geburtstag [der kleine-n Lotte]$_{POSSESSOR}$* 'the birthday of little Lotty'). Functionally, the PROP-genitive is limited to possession. Prepositions which govern the genitive on COMMs as their complement do not trigger the s-genitive on PROPs. Similarly, verbs which govern the genitive of COMMs as objects allow for prepositional objects with PROPs.

The German facts speak in favour of SOG. Not so much so in word morphology since s-genitives are also mandatory for masculine and neuter loan nouns ending in a (full) vowel other than schwa such as *die Windschutzscheibe [des Auto-s]$_{POSSESSOR}$* 'the windscreen of the car'. The situation is clearer beyond the word-boundary. The single marking (termed *Monoflexion* by Nübling et al. (2015: 68)) of the genitive in phrases containing a PROP as opposed to the multiple marking of the same category with COMMs in the identical syntactic context counts as a SOG-property (Spencer 2007: 183–190). On this basis, we venture to propose the split rule in Figure 8 for German.

---

44 What is illustrated with ANTHs in this and the subsequent paragraph also holds for TOPOs in German. Thus, we can speak of SOG.
45 Eisenbeiß et al. (2009: 158–159) mention that the term *proper name possessive marker* is used by some linguists of German for the s-genitive of PROPs.

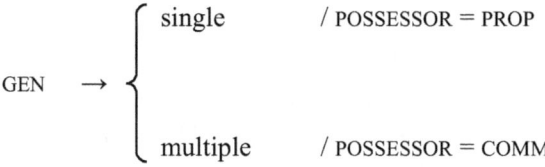

Figure 8: Split rule in German I.

Much the same holds for the word-order differences reported above – or so it seems. However, the additional criterion of synctatic weight must be integrated into the rule. This criterion causes a division of PROP-possessors in two with light ones being admitted to the slot on the left of the possessee and heavy ones being positioned on the right. Since these observations do not yield strict rules but capture tendencies and preferences, the split rule in Figure 9 must be taken with a grain of salt.

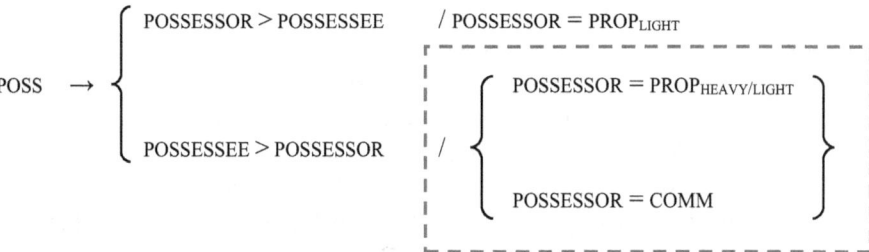

Figure 9: Split rule in German II.

The zone of overlap of PROPs and COMMs is relatively large. The present state is most probably only transitory in the sense that in times to come all possessors – light or heavy, PROP or COMM – might wind up in the same position on the right of the possessee. Synchronically, there undeniably is relatively robust evidence of SOG in German – and this means standard German.

A close relative of German – Dutch – has lost its former case inflections almost completely. There is the residual genitive about which Booij (2002: 35) states that

> [t]he only nouns that can be used with this kind of possessor marker are proper names and nouns that can be used as forms of address [. . .], that is, words functioning as proper names with an inherent referential value, and some pronouns.

We are facing another case of SAG of Type IV. The special behaviour of the Dutch genitive is restricted to ANTH(-N)s (36a). If COMMs are adnominal possessors the employment of the preposition *van* 'of' is obligatory (36c). This preposition is also compatible with PROP-possessors (36b) (cf. Stolz et al. (2008: 390–396)).

(36) Dutch [Indo-European, Germanic]
    (a) possessor = ANTH I {HP I Dutch, 48}
        het   leek   alsof   [[Hagrid-s]$_{POSSESSOR}$ jas]   alleen maar
        it   seem.PST as_if [[Hagrid-GEN]   jacket] alone   but
        uit   zakken   bestond
        out   pocket:PL consist.PST
        '[[Hagrid's] coat] seemed to be made of nothing but pockets [. . .].'
        {HP I English, 72}
    (b) possessor = ANTH II {HP I Dutch, 75}
        hij   vertelde Ron dat   hij   altijd   [de   oude   kleren
        3SG.M tell:PST   Ron that   3SG.M always [DEF.PL old   clothe:PL
        [van   Dirk]$_{POSSESSOR}$] had   moeten   dragen
        [of   Dirk]]   have.PST must:INF wear:INF
        '[. . .] he told Ron [. . .] all about having to wear [[Dudley's] old clothes]
        [. . .].' {HP I English, 111}
    (c) possessor = COMM {HP I Dutch, 64}
        en   uiteraard bereikt   u   nooit zo   'n   goed   resultaat
        and   of_course reach:2SG HON   never such INDEF good   result
        met   [de   stok   [van een   andere   tovenaar]$_{POSSESSOR}$]
        with [DEF.UT wand   [of   INDEF other   magician]]
        'And of course, you will never get such good results with [[another
        wizard's] wand].' {HP I English, 95}

As in the German case discussed above, there is some overlap between ANTHS and COMMS since both are entitled to make use of the *van*-construction. However, COMMS are not allowed to take the *s*-genitive which remains the monopoly of ANTHS. Thus, Dutch attests to SAG of Type IV. Figure 10 reveals that, in contrast to COMMS, ANTHS are structurally privileged insofar as they have one more option to choose from.

$$\text{POSS} \rightarrow \begin{cases} \text{-}s & /\text{ POSSESSOR} = \text{ANTH} \\ \\ van & /\text{ POSSESSOR} = \begin{Bmatrix} \text{ANTH} \\ \text{COMM} \end{Bmatrix} \end{cases}$$

Figure 10: Split rule in Dutch.

In spite of the possibility of ANTHs and COMMs taking the *van*-construction, there remains an area in which ANTHs are without competition from another nouny word-class.

For **West Frisian** [Indo-European, Germanic], Popkema (2018: 151–152) and Hoekstra (2018: 40–42) report on the co-existence of two inflectional genitives, both of which are blocked for COMMs. We illustrate the situations with examples from the former. There is the productive *s*-genitive which is attached to PROP(-N)s as in *Klaske-s fyts* 'Klaske's bicycle', *Fryslân-s verlede* 'Frisia's past', *mame-s bok* 'mother's book', etc. The second inflectional genitive ending *-e* is no longer productive and restricted to ANTH(-N)s (including kinship terms but not TOPOS) as in *Jann-e Teats* 'Jan's Teats' (*Teats* being a female ANTH), *heit-e kleren* 'father's clothes', etc. Popkema (2018: 151) claims that for all other kinds of possessor nouns (= COMMs) the preposition *fan* 'of' serves as relator as in *de tsjillen fan de karre* 'the wheels of the car'. The descriptive grammar of West Frisian keeps silent about two facts. First, the *fan*-construction is also very common with PROPs as possessors as shown in (37a). Second, there is a third and frequently attested construction-type in adnominal possession which seems to be possible only with possessors which belong to the PROP-class, namely the resumptive-pronoun construction illustrated in (37b).

(37) West Frisian [Indo-European, Germanic] {HP I West Frisian, 48}
    (a) *fan*-construction
        De   ûle   fladdere   op   'e   grûn   del   en   pikte
        DEF.UT owl   flutter:PST on   DEF.UT   ground   down   and   pick:PST
        nei   [de   jas   *[fan   Hagrid]*_{POSSESSOR}]
        to   [DEF.UT jacket [of   Hagrid]]
        'The owl then fluttered on to the floor and began to attack [[Hagrid's] coat].' {HP I English, 71–72}
    (b) resumptive pronoun
        [*Hagrid*]_{POSSESSOR}   **syn**   jas]   wie   allinne   mar
        [Hagrid]   POSS.3SG.M   jacket]   be.3SG.PST   alone   but
        bûsen
        pocket:PL
        '[[Hagrid's] coat] seemed to be made of nothing but pockets [...].' {HP I English, 72}
        [lit.] '[[Hagrid's] coat] was made of nothing but pockets [...].'

The resumptive-possessive construction will be discussed in more detail in the subsequent paragraph on Low German. At this point, it suffices to conclude that in West Frisian adnominal possession, COMMs as possessors must take the *fan*-construction whereas PROPs and especially ANTHs can choose from several options.

Three of these options are exclusive to PROPS (or ANTHS). It is of no relevance that one of these SAG-traits is unproductive. As Figure 11 shows there is evidence of SOG and SAG in West Frisian. Since the identical or differential behaviour of TOPOS and ANTHS is construction-specific, we cannot assign West Frisian to one particular type alone. There is SOG – but the picture is heterogeneous.

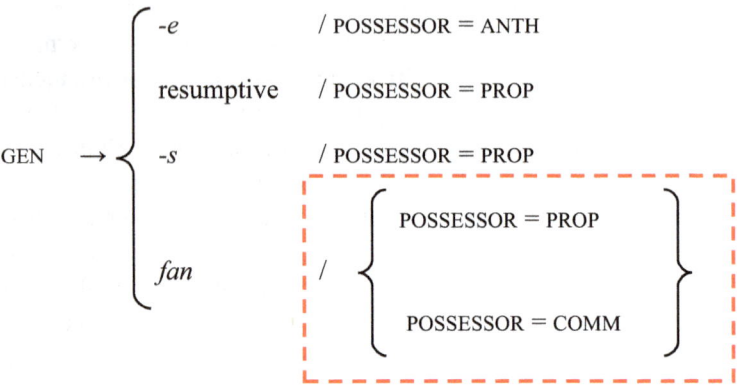

Figure 11: Split rule in West Frisian.

There is an overlap between COMMS and PROPS but at the same time, there are constructions which are not accessible to COMMS.

The resumptive-possessive construction [POSSESSOR **POSSESSIVE PRONOUN** POSSESSEE] is attested in many Germanic varieties (Poulsen 2002: 39; Braunmüller 2018: 315–318). Booij (2002: 35) mentions this construction type for Dutch with the possessive pronoun of the 3rd person singular (full forms M.SG *zijn* 'his' and F.SG *haar* 'her', reduced *z'n* and *d'r*, respectively). That this construction type is at home in the spoken register is irrelevant for the point we are going to make. Examples like *de directeur **z'n** kamer* 'the director's room' are possible with ANTHS and COMMS alike. The only criterion the possessor has to meet is that of animacy. In contrast to Dutch, some Germanic languages (or their nonstandard varieties) do not tolerate possessors other than PROPS in the resumptive-possessive construction. This is the case in **Low German** [Indo-European, Germanic], for instance. Lindow et al. (1998: 144) state that the erstwhile inflectional genitive is replaced with either the *vun*-construction (involving the preposition *vun* 'of') or the resumptive-possessive construction. The authors do not mention any criteria which determine the choice of construction. However, the examples they provide are indicative of a distribution over PROPS and COMMS which resembles that identified for West Frisian above. The NP *[[den Vadder]$_{POSSESSOR}$ **sien** Huus]* 'father's house' illustrates the resumptive-possessive construction whereas the NP *[dat Dack [**vun** den Schuppen]$_{POSSESSOR}$]* 'the roof

of the shed' is given for the *vun*-construction. The possessor in the resumptive-possessive construction is a kinship term. In the *vun*-construction, the possessor is an inanimate COMM. One might assume that the former construction is restricted to ANTH(-N)S. However, the Low German case is not straightforward as results from the findings of Stolz et al. (2008: 375–390). In (38), we present examples of the resumptive-possessive construction with an ANTH-possessor in the masculine singular (38a), an ANTH-possessor in the feminine singular (38b), a coordinated ANTH-possessor with plural agreement on the pronoun (38c), the *vun*-construction with an ANTH-possessor (38d), the *vun*-construction with a COMM-possessor, and (38f) the resumptive-possessive construction with a COMM-possessor.

(38) Low German [Indo-European, Germanic]
    (a) resumptive, possessor = ANTH M.SG {HP I Low German, 332}
        dat   weer   de   beste   Avend   in   [Harry*POSSESSOR*
        DEM.DIST be.PST.3SG DEF.M best   evening in   [Harry
        [sien   *Leven]]*
        [POSS.3SG.M life]]
        'It was the best evening in [[Harry's] life] [. . .].' {HP I English, 330}
    (b) resumptive, possessor = ANTH F.SG {HP I Low German, 321}
        hebbt   Se   [Hermine*POSSESSOR* [ehr   Uhl]]   kregen?
        have.2SG HON   [Hermine   [**POSS.3SG.F** owl]] receive:PTCPL
        'You got [[Hermione's] owl]?' {HP I English, 319}
    (c) resumptive, possessor = ANTH PL {HP I Low German, 17}
        he   hett   versöcht   [Potters*POSSESSOR* [ehrn
        3SG.M have.PST.3SG try:PTCPL   [Potter:PL   [**POSS.3PL.ACC**
        Söhn   Harry]]   ümtobringen
        son   Harry]]   kill:PTCPL
        '[. . .] he tried to kill [[the Potters'] son].' {HP I English, 19}
    (d) *vun*-construction, possessor = ANTH {HP I Low German, 57}
        ook   'n   Mann mit   mehr   Kraasch as   Unkel Vernon
        also INDEF   man   with   more   power   than   Uncle Vernon
        weer   nu   vör   [dat   füünsch Gesicht *[vun*
        be.PST.3SG now for   [DEF.NT angry   face   [of
        Hagrid]*POSSESSOR*] tohoopklappt
        Hagrid]]   collapse:PTCPL
        'A braver man than Vernon Dudley would have quailed under the furious look Hagrid now gave him [. . .].' {HP I English, 59}
        (lit.) 'Even a man with more power than Uncle Vernon would now have collapsed because of [the angry face [**of** Hagrid]].'

(e) *vun*-construction, possessor = COMM {HP I Low German, 326}
he   is   jo   ook [de   Baas [vun de
3SG.M be.3SG.PRS well also [DEF.M boss [of   DEF.F
School]$_{POSSESSOR}$]
school]]
'Well, of course, that was the Headmaster [...].' {HP I English, 323}
(lit.) '[...] he is of course [the director [**of** the school]].'

(f) resumptive, possessor = COMM {HP I Low German, 299}
Ron   beet   de   Tähn   tohoop   un   stapp
Ron bite.PST.3SG DEF.PL tooth.PL together and step.PST.3SG
vörsichtig över [den   Hund$_{POSSESSOR}$ [sien   Been]] weg
carefully   over [DEF.M.ACC dog   [POSS.3SG.M leg]]   away
'Ron gritted his teeth and stepped carefully over [[the dog's] legs].'
{HP I English, 297}

Superficially, it does not seem to matter whether the possessor is an ANTH or a COMM because both construction types are possible with either word-class. On closer inspection, the picture changes considerably because there are clear preferences. As Stolz et al. (2008: 382–383) observe with reference to the corpus text HP I Low German, all possessors in HP I Low German in resumptive-possessive constructions are animate whereas three-quarters of all possessors in *vun*-constructions are inanimate. What is even more striking is the preponderance of ANTHs among the possessors in resumptive-possessive constructions and their small turnout in the competing construction. The discrepancy comes to the fore in Figure 12.

ANTHs account for 88% of all possessors in resumptive-possessive constructions[46] but their share is down to 6% in the case of *vun*-constructions. How do we explain the latter cases? Of the nineteen tokens of ANTH-possessors in the *vun*-constructions, eleven (~ 58%) involve syntactically heavy NPs as possessees, i.e., the NP contains modifiers such as attributive adjectives and/or further determiners. The share of resumptive-possessive constructions which involve both an ANTH-possessor and a syntactically heavy possessee amounts to only 16% as shown in Figure 13. The quantities tell the following story. The *vun*-construction is an option for ANTH-possessors especially if the possessee is a multi-word syntagma. Figure 14 shows that the situation is similar for heavy possessors, meaning: ANTHs which consist of more than just one word unit have a greater share of the *vun*-constructions than they have of the resumptive-possessive constructions.

---

[46] Since the resumptive-possessive construction is also possible with pronominal possessors (six tokens in the corpus text), the share of ANTHs increases to 90%.

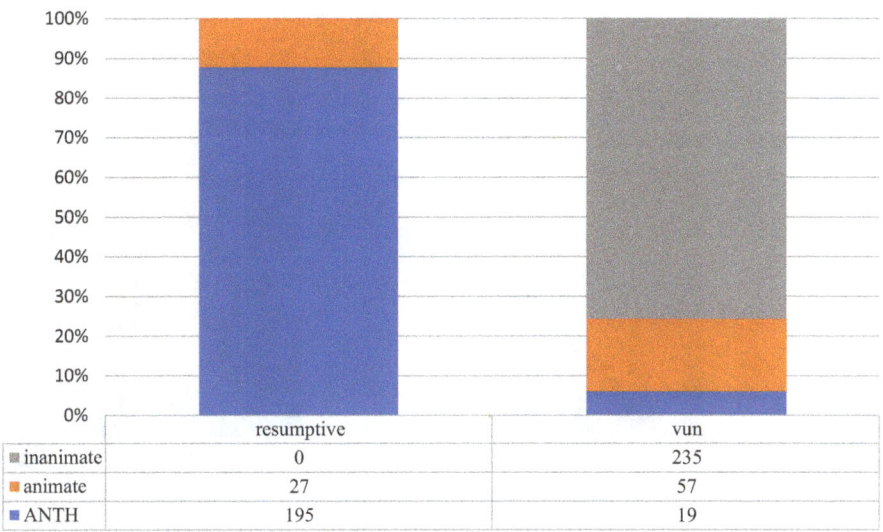

Figure 12: Distribution of possessors over constructions in Low German.

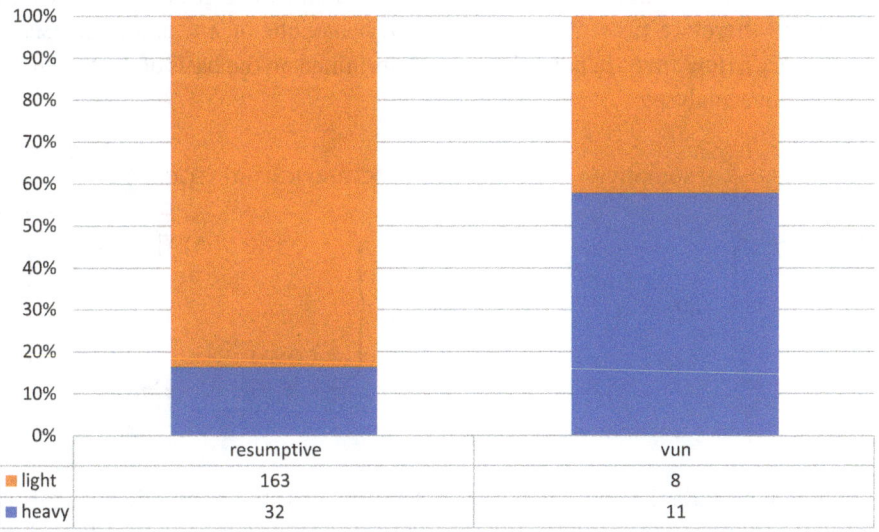

Figure 13: Heavy vs light possessee NPs over construction types in Low German.

Stolz et al. (2008: 389) also discuss focus and indefiniteness as factors which determine the choice of construction with ANTH-possessors. According to their hierarchy of factors, an adnominal possessive relation tends to be expressed by the resumptive-possessive construction if the possessor is an ANTH in a syntactically light

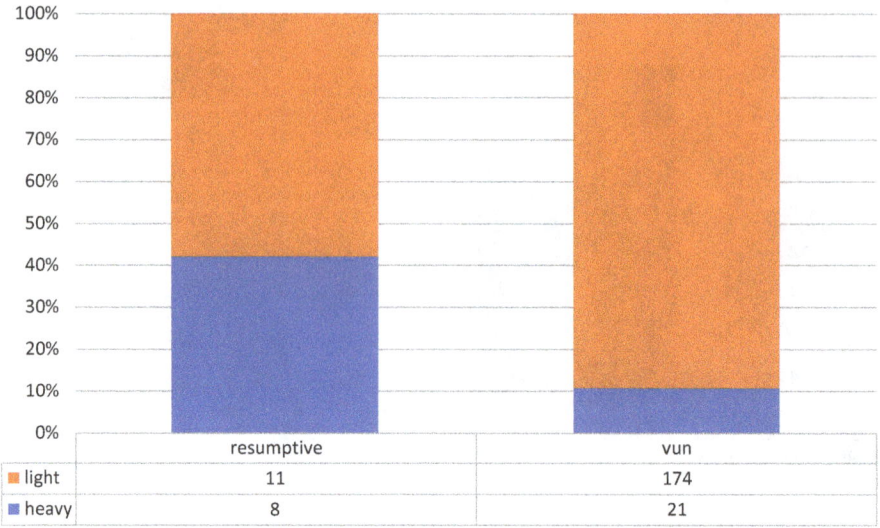

	resumptive	vun
light	11	174
heavy	8	21

Figure 14: Heavy vs light possessor NPs (ANTHS) over construction types in Low German.

out-of-focus NP. It is not possible to postulate hard and fast rules for Low German. Figure 15 therefore presents the differential behaviour of ANTHS and COMMS as preferences. How TOPOS behave cannot be determined on the basis of the text material we have analysed.

$$\text{POSS} \rightarrow \begin{cases} \text{resumptive} & / \text{ POSSESSOR} = \text{ANTH\&LIGHT\&[-FOCUS]} \\ \textit{vun} & / \text{ POSSESSOR} = \begin{cases} \text{ANTH\&HEAVY[+FOCUS]} \\ \text{COMM} \end{cases} \end{cases}$$

Figure 15: Opposing tendencies in Low German.

We do not exclude the possibility that in-depth corpus studies on other languages might yield a similar result in the sense that the supposedly clearcut boundary between COMMS and PROPS is fuzzy because factors other than word-class membership come into play, too.

4.2.2 A genitive only for names: Faroese

A by far more convincing SAG-case is **Faroese** [Indo-European, Germanic]. This insular North Germanic language displays a particularly rich and differentiated system of adnominal and pronominal possessive constructions (Stolz and Gorsemann 2001). The choice of construction is determined on the one hand by the semantics of the possessee – a situation that is widely common across languages. However, the possessor also has a say when it comes to choosing the appropriate possessive construction. In this context, it is important to understand that the inflectional genitive of nouns is largely obsolete. Thráinsson et al. (2004: 61) claim that of the four inherited morphological cases nominative, accusative, dative, and genitive, the latter has become unproductive (Petersen and Szczepaniak 2018: 115–117). Somehow, however, inflectional genitives are still around because

> the genitive form of many nouns occurs in fixed expressions and also as first part of certain compounds (although it is doubtful that speakers intuitively interpret such forms (parts of words) as a genitive [. . .]). [G]enitive forms [. . .] might be found in written texts, even though they are not common in the modern spoken language. [T]he adnominal (or "possessive") genitive is very unnatural in most of these cases [of adnominal possession] in Modern Faroese. [. . .] Genitive forms of nouns are used with some prepositions, especially in fixed expressions. (Thráinsson et al. 2004: 62–63)

This quote describes the inflectional genitive as being increasingly marginalised in Faroese where it still occupies small functional niches with strong associations to the written register. In (39), we provide an example of the still relatively frequent employment of the inflectional genitive required by the directional preposition *til* 'to'.

(39) Faroese [Indo-European, Germanic] {Dir 9}
 maðurin   skal   til   Frakland-s   at   arbeiða   í   tveir
 man:NOM:DEF shall to   France-GEN   to   work:INF in two:ACC
 mánaðir
 month:ACC.PL
 '[. . .] the husband shall go to France to work for two months (there) [. . .].'

The preservation of the inflectional genitive is a characteristic trait of TOPOS although "[a] few place names do not use *til* to express motion towards [. . .] and hence have [...] no genitive at all" (Lockwood 1977: 38).⁴⁷ *Til* 'to' governs the accusative with definite or modified COMMs (Thráinsson et al. 2004: 179).

Lockwood (1977: 28) classifies the residual strongholds of the genitive as "Icelandicisms", i.e., as artificial uses resulting from the efforts of some Faroese writers to revitalise the already dead genitive on the pattern of Icelandic where the inflectional genitive is still very much alive. The obituary of the erstwhile genitive raises the question of how adnominal possessive relations are expressed now that the old strategy is no longer an option. To start with, there are three prepositional replacements for the inflectional genitive. The three prepositions have a basic spatial meaning and govern the dative on the possessor. As the default, the preposition *hjá* 'at' may neutralise all other distinctions in the possessive domain with focus, (in) definiteness, and/or syntactic weight of either possessor or possessee often coming into play in favour of the choice of *hjá* 'at'. In (40), the use of the default preposition with a COMM (40a) and an ANTH (40b) as possessor is illustrated.

(40)　Faroese [Indo-European, Germanic]
　　(a)　*hjá* + COMM (Thráinsson et al. 2004: 62)
　　　　her　　eru　　 [húsini　　　　[hjá　einum　　　ríkum　　manni]$_{POSSESSOR}$]
　　　　here　be:3PL　[house:PL:DEF　[at　　INDEF:DAT　rich:DAT　man:DAT]]
　　　　'Here is a [[a rich man's] house/home].' [O.T.]
　　(b)　*hjá* + ANTH {Dir 20}⁴⁸
　　　　hon　　 tók　　　　　 í　　[stóru　　　taskuna　　　　　　[hjá　Dirdri]$_{POSSESSOR}$]
　　　　3SG.F　take.PST.3SG　in　[big:ACC　bag:ACC:DEF:ACC　[at　　Dirdri.DAT]]
　　　　'She took [Dirdri's [big bag]].'

The two prepositions *í* 'in' and *á* 'on' display a peculiar distribution. If the possessor is pronominal or a human noun, only inalienable possessees are admitted to the construction – in this case, body-part terms and physico-mental states. If the possessor is a COMM, however, further part-whole relations and functions in a given institution are covered, too. We exemplify the use of these prepositions only with body-part possession in (41) for COMM-possessors and (42) for ANTH-possessors.

---

47 The preservation of the inflectional genitive with most but not all TOPOs as well as the use of prepositions other than *til* to express Goal with certain TOPOs (Lockwood 1977: 96–97) suggest that there might be reason to postulate the existence of STG in Faroese, too. For the purpose of this section, we skip this issue since the focus is on SAG.
48 The ANTH *Dirdri* is indeclinable so that the dative is formally indistinguishable from the other cases.

(41) Faroese [Indo-European, Germanic]
   (a) *á* + COMM {HP I Faroese, 11}
   [*halin*   [*á*   *kettuni*_POSSESSOR_]   *livdi*
   [tail:DEF   [on   cat:DEF:DAT]]   live:PST.3SG
   'The [[cat's] tail] twitched [...].' {HP I English 15}
   (b) *í* + COMM {HP I Faroese, 12}
   hetta   voru   [*eyguni*   [*í*   *kettuni*_POSSESSOR_]
   DEM   be.PST.3PL   [eye:PL:DEF   [in   cat:DEF:DAT]]
   '[...] which were [the eyes [of the cat]] [...].' {HP I English 15}

(42) Faroese [Indo-European, Germanic]
   (a) *á* + ANTH {Dir 82}
   okkurt   trýsti   um   [*hálsin*   [*á*   *Dirdri*_POSSESSOR_]
   something   press:PST.3SG   about   [neck:DEF.ACC   [on   Dirdri.DAT]]
   'Something pressed against [[Dirdri's] neck].'
   (b) *í* + ANTH {Sum 30}
   Jóannes   sá   [*eyguni*   [*í*   *Trygv-a*_POSSESSOR_]   *í*
   Jóannes   see.PST.3SG   [eye:PL:DEF   [in   Trygvi-DAT]]   in
   bakkspeglinum
   rearview_mirror:DEF.DAT
   'Johannes saw [[Trygvi's] eyes] in the rearview mirror.' (Stolz and Gorsemann 2001: 581)

Superficially, these examples seem to indicate that the word-class membership of the possessor is largely irrelevant provided it is a noun. This supposed irrelevance does not apply in other domains though. Possessors have to be animate – and thus are frequently ANTHs – in a construction type which is specialised for kinship relations. In this construction type, the possessor is in the accusative (Lockwood 1977: 103; Thráinsson et al. 2004: 33) as shown in (43).

(43) Faroese [Indo-European, Germanic]
   (a) possessor = COMM {Reg 50}
   [*kon-a*   [*stjór-a-n*_POSSESSOR_]   *tók*   *telefonina*
   [woman-NOM   [director-ACC-DEF]]   take.PST.3SG   telephone:DEF:ACC
   '[[The director's] wife picked up the phone.' (Stolz and Gorsemann 2001: 567)

(b) possessor = ANTH {Dir 50}⁴⁹
    Gumman   [mamma   [Jaspur   og   Fí-u]_{POSSESSOR}]
    godmother:NOM:DEF [mother.NOM [Jaspur.ACC and Fía-ACC]
    vendi   sær   móti   henni
    turn:PST.3SG RFL.DAT against 3SG.F.DAT
    'The godmother, [the mother [of Jaspur and Fía]], turned towards her.'

The behaviour of ANTHs in (40b), (42), and (43b) is not unique to them. They share it either with COMMs in general or with human nouns. So far, it is not possible to speak of SAG in the realm of possession.⁵⁰ However, ANTHs have still one more property in store which none of the other nouny word-classes have access to. This property is the *sa*-genitive (aka *sa*-possessive (Staksberg 1996)).

The *sa*-genitive has been discussed repeatedly in the literature (Staksberg 1996, Thráinsson et al. 2004: 248–252, Stolz et al. 2008: 220–221, Stolz et al. 2017a: 125–126, Braunmüller 2018: 309–310). Lockwood (1977: 106) states that

> [a]fter names, or nouns used as such, the spoken language commonly forms a genitive of possession with the suffix *-sa(r)*, though the literary language mostly avoids this construction.

The scarcity of *sa*-genitives in written Faroese is still a fact as will be argued below. Staksberg (1996: 30) disagrees with Lockwood as to the exact morphological status of the marker which to his mind is a clitic *=sa* and not an inflectional suffix. This analysis is accepted by Thráinsson et al. (2004: 64) who further claim that

> [t]his marker cannot be suffixed to common nouns nor to place names and it cannot be governed by prepositions that can govern the regular morphological genitive.

The examples in (44) are the only attestations of the *sa*-genitive we have come across in our corpus of literary Faroese. The *sa*-genitive is glossed as GEN2 to distinguish it from the obsolete GEN1 to which the clitic is attached.

---

49 The ANTH *Jaspur* does not formally distinguish between nominative and accusative since the final two segments /ur/ belong to the stem.

50 It should not go unmentioned that Faroese gives evidence also of an associative plural which is restricted to ANTH(-N)S. The construction type is [ANTH 3PL], i.e., the ANTH is directly followed by the personal pronoun of the 3ʳᵈ person plural in the appropriate case as shown by underlining in (i).

(i) Faroese [Indo-European, Germanic] {Dir 84}
    Inga   var   eisini uppi í [túninum   [hja Dirdri og teimum]_{POSSESSOR}]
    Inga   be.PST.3SG still   up   in [garden:DEF:DAT [at Dirdri and 3PL:DAT]]
    Inga was still in [the garden [of Dirdri and family]].'

(44) Faroese [Indo-European, Germanic]
   (a) possessor = ANTH {Dir 65}[51]

áðrenn	hon	var	komin	yvir	um	hondina
before	3SG.F	be.PST.3SG	come:PTCPL	over	about	hand:DEF:ACC

hjá	Dirdri	hevði	Kári	lagt	sína	hond
at	Dirdri.DAT	have:PST.3SG	Kári	lay:PST.3SG	POSS.3SG:ACC	hand

hinumegin	[[Dirdri=**sa**]$_{POSSESSOR}$	___]
other_side	[[Dirdri=**GEN2**]	___]

   'Before it had come over Dirdri's hand, Kári had laid his hand on the other side of [[Dirdri's] (hand)].'

   (b) possessor = ANTH {Dir 65}

so	fór	svartaklukkan	spákandi	víðari	yvir
then	go.PST.3SG	beetle:NOM:DEF	go_for_stroll:PTCPL	further	over

á	[[Kár-a=**sa**]$_{POSSESSOR}$	hond]
on	[[Kári-GEN1=**GEN2**]	hand]

   'Then the beetle crawled further onto [[Kári's] hand].' (Stolz and Gorsemann 2001: 567)

   (c) possessor = kinship term (ANTH-N){Dir 84}

kuffertið	og	stóra	taskan	stóðu	úti	á
suitcase:DEF.NT	and	big	bag:NOM:DEF	stand.PST:3PL	outside	on

[[omm-u=**sa**]$_{POSSESSOR}$	trappu]
[[grandma-GEN1=**GEN2**]	doorstep:DAT]

   'The suitcase and the big bag stood outside on [[grandma's] doorstep].'

In (44a), the possessee *hond* 'hand' has fallen victim to equi-deletion. Adverbs ending in *-megin* like *hinumegin* '(on) the other side of' can be used as prepositions which govern the accusative (Lockwood 1977: 91). Thus, *Dirdrisa* cannot be the complement of *hinumegin*. Sentences (44a) and (44b) follow each other directly in this order in the original story. It is clear therefore that the girls let the beetle crawl from the one's hand over to the hand of the other. As to the role of GEN1 in the formation of the *sa*-genitive, Thráinsson et al. (2004: 64) argue that relics of the erstwhile inflectional genitive (or non-nominative) are required on so-called weak nouns (= ANTHs like *Beint-a* → *Beint-u=sa*) to host the *sa*-clitic whereas ANTHs which inflect according to the strong declensions tend to attach the *sa*-clitic to the nominative (e.g. *Sjúð-ur* → *Sjúð-ur=sa* ~ *Sjúð-a=sa*). Moreover, the *=sa* is a phrasal clitic which takes the rightmost constituent of the NP as its host. This entails that under

---

[51] The sentence-final underlined empty slot in (44a) indicates where the equi-deleted COMM *hond* 'hand' would be placed.

coordination the clitic is attached only to the final conjunct whereas the previous conjuncts inflect for the GEN1 provided the ANTH is of the weak-noun type as in *Beint-u og Ro-a=sa bók* {Beinta}-{GEN1} {and} {Rói}-{GEN1}={GEN2} {book} 'Beinta and Roi's book' (Thráinsson et al. 2004: 64). In a way, the *sa*-clitic provides another niche for the survival of the GEN1. =*sa* cliticises also to multi-word NPs such as
- *Tummas á Dómarakontór-i-n-um=sa bil-ur* {Thomas} {on} {legal_offic}-{DAT}-{DEF}-{DAT}={GEN2} {car}-{NOM} 'Thomas at the legal office's car' (Thráinsson et al. 2004: 64)

where the host of the clitic is the case-inflected complement of a preposition which in turn is the head of the prepositional attribute of the ANTH *Tummas*.

In Table 10, the case inflection in the singular of the COMMS *tunga* 'tongue' and *fuglur* 'bird' are compared to those of the ANTHs *Beinta* and *Sjúður*.

Table 10: Paradigms of COMMS and ANTHS compared.

case	weak		strong	
	COMM	ANTH	COMM	ANTH
NOM	*tung-a*	*Beint-a*	*fugl-ur*	*Sjúð-ur*
ACC	*tung-u*	*Beint-u*	*fugl*	*Sjúð-a*
DAT	*tung-u*	*Beint-u*	*fugl-i*	*Sjúð-i*
GEN	*(tung-u)*	*Beint-u=sa*	*(fugl-s)*	*Sjúð-ur=sa ~ Sjúð-a=sa*

The grey shading identifies the area in which ANTHs and COMMs go separate ways morphologically. ANTHs differ from COMMs in multiple ways.
- While the inflectional genitive is on its way out of the paradigm of COMMs, it is preserved to some extent with ANTHs because, for weak declension ANTHs at least, it is the form to which the *sa*-clitic is attached.
- The innovative *sa*-genitive requires the linearisation POSSESSOR > POSSESSEE (e.g. *Ólav(-ur)=sa bilur* 'Olavur's car') whereas the old genitive would be possible only with the inverse order POSSESSEE > POSSESSOR (e.g. *bilurin Ólav-s* 'Olavur's car') (Thráinsson et al. 2004: 249).[52]

---

52 Thráinsson et al. (2004: 249) mark the latter example as doubtful. This is probably caused by the absence of the definiteness marker on the possessee *bilur* 'car' in the original. We have taken the liberty to adjust the example accordingly.

- The *sa*-genitive is functionally limited to expressing adnominal possession. This is exactly the domain from which the inherited GEN1 has disappeared in the grammar of COMMS.[53]
- GEN1 is a bona fide representative of bound morphology whereas the *sa*-genitive belongs to a different morpheme class, namely that of phrasal clitics.

The *sa*-genitive is functionally and formally special within Faroese morphology. It is responsible for an array of structural mismatches in the sense that it does not conform to the canonical patterns which dominate throughout the declension system of nouns. Add to this the use of the prepositions *at* 'at, to' (+ DATIVE) and *til* 'to' (+ ACCUSATIVE) "with names and terms of relationship" (Lockwood 1977: 105) as in *omma at Marjuni* 'Marjun's grandmother' (Thráinsson et al. 2004: 160) and *mamma til Líggjas* 'Elijah's mother' (Lockwood 1977: 105) – two phenomena which call for further scrutiny in a separate study – and the pervasiveness of SAG in the domain of possession in Faroese can be claimed to be established beyond doubt.

Given the many adnominal constructions which co-exist in Faroese, Figure 16 necessarily simplifies the intricate system to emphasise the separation of ANTHS from COMMS.

$$\text{POSS} \rightarrow \begin{cases} =sa & / \text{ POSSESSOR = ANTH(-N)} \\ at \sim til & / \text{ POSSESSOR = ANTH-N \& POSSESSEE = ANTH} \\ \text{ACC} & / \text{ POSSESSOR = ANTH-N \& POSSESSEE = HUMAN} \\ i \sim á & / \text{ POSSESSOR = X \& POSSESSEE = BODY-PART} \\ hjá & / \text{ POSSESSOR = X} \end{cases}$$

Figure 16: Split rule in Faroese.

There is again a zone of overlap where both ANTHS and COMMS behave identically. However, beyond the limits of this zone, there is a domain in which only ANTH(-N)S

---

53 An alternative account which we cannot elaborate upon in this study assumes that =*sa* is an enclitic postposition which governs the GEN1 (or under the appropriate phonological conditions the nominative(!)) on the ANTH. Even under this alternative account, =*sa* creates a mismatch since as a postposition it would be the sole representative of its kind in an otherwise prepositional language (Thráinsson et al. 2004: 151–152).

are allowed to function as either possessor or possessee. This means that Faroese attests to SAG of Type IV. Section 4.3 will answer the question of whether the closest relative of Faroese, Icelandic, behaves similarly.

4.3 Against parsimony: The pronominal possessive in Icelandic

In his discussion of the *sa*-genitive in Faroese, Staksberg (1996: 33) very briefly refers to the resumptive-possessive construction in Norwegian as a kind of distant parallel. Delsing (1993: 147–185) provides a pan-North Germanic account of adnominal possession which also features the resumptive-possessive construction. More importantly, the author observes that

> the [Scandinavian] languages differ considerably. There are often different constructions depending on the type of possessor used: possessive pronoun, personal name or ordinary noun phrase. Some constructions are only found with personal names, whereas the ones that are applicable to ordinary DPs are also possible with personal names (Delsing 1993: 158–159).

In this section, we focus on the proprial possessive construction which is the monopoly of ANTH(-N)-possessors. Delsing (1993: 53–55) shows that in several north Germanic languages – mostly in nonstandard varieties or colloquial registers – ANTHs are accompanied by the appropriate personal pronoun of the 3rd person singular (mostly when used as arguments). A pertinent Norwegian example is given in (45).

(45) Norwegian [Indo-European, Germanic] (Delsing 1993: 54)
     han   Per har   slage **ho**   Kari
     3SG.M Per have:PRS hit 3SG.F Kari
     'Per has hit Kari.'

The pronouns agree in gender with the referent of the ANTH they accompany. Normally, these prenominal pronouns are insensitive to case (Delsing 1993: 55, footnote 35). The prenominal pronouns as such call for being studied in-depth separately – a task we relegate to future work of ours. What is interesting for the discussion in this section, however, is the frequent use of prenominal pronouns in adnominal possessive constructions. Delsing (1993: 157) considers this usage to be typical of colloquial **Icelandic** (where cases like those in (45) are also attested with a "special stylistic value" (Thráinsson 2010: 91)).[54] The phenomenon goes by the

---

[54] Comprehensive accounts of possession in Icelandic can be found in Stolz (2004), Stolz et al. (2008), Thráinsson (2010), Stolz (2012), and Schuster (2020).

name of proprial possessive construction and has already been addressed repeatedly for instance by Stolz et al. (2008: 139–142) and Stolz et al. (2017a: 126–128).[55] Alternatively, the pronoun employed in this construction is called proprial article (Thráinsson 2010: 91).

In the domain of adnominal possession, Icelandic resembles Faroese insofar as the prepositions *í* 'in' and *á* 'on' introduce the possessor in constructions which involve inalienable possessees such as body-parts (Stolz et al. 2008: 142–144). Compare (46) to (41)–(42) above.

(46) Icelandic [Indo-European, Germanic]
    (a) *á* + POSSESSOR {Jón 32}
        [*stóra   táin*   [*á*   *pabba*]$_{POSSESSOR}$]   *er*   *eins*
        [big:NOM toe:DEF [on father:DAT]] be.3SG one:GEN
        og   *apríkósa*
        and   apricot:NOM
        '[[Father's] big toe] is like an apricot.' (Stolz et al. 2008: 143)
    (b) *í* + POSSESSOR {Jón 20}
        [*annað   augað*   [*í*   *Jóni*   *Bjarna*]$_{POSSESSOR}$] *rann*
        [other:NT eye:NOM:DEF.NT [in Jón:DAT Bjarni:DAT]] run.PST
        alveg   inn að nefinu
        always in to nose:DEF:DAT
        '[[Jón Bjarni's] other eye] always looked towards the nose.' (Stolz et al. 2008: 144)

The differences between the two insular North Germanic languages cover the productivity of the inherited inflectional genitive in Icelandic as opposed to the loss of this case and its renewal with ANTHs in Faroese, the possibility of possessors taking the accusative in Faroese, and the ubiquitous preposition *hjá* 'at' for alienable possession in Faroese (Stolz et al. 2008: 221). What additionally distinguishes Icelandic from Faroese is the predilection of the former for the proprial possessive construction which seems to be alien to Faroese.

The proprial possessive construction has the following shape:
- [COMM$_{POSSESSEE}$ [PRONOUN$_{GEN}$ ANTH$_{GEN}$]$_{POSSESSOR}$],

---

[55] In her detailed diachronic study of Icelandic possession from the earliest texts to the present day, Schuster (2020: 243–244) explicitly skips the proprial possessive construction because of its supposedly very restricted functions.

i.e., the possessor consists of the sequence of the genitive form of the personal pronoun of the 3ʳᵈ person and the coreferential ANTH inflected in the genitive. The pronoun corresponds formally to the gender of the ANTH as shown in (47).

(47) Icelandic [Indo-European, Germanic]
    (a) masculine possessor {HP I Icelandic, 227}
        hvar   er   [skrifstofan   **[hans**   Dumbledore-s]$_{POSSESSOR}$]
        where be.3SG [office:DEF   [3SG.M:GEN Dumbledore-GEN]]
        'Where's [[Dumbledore's] office]?' {HP I English, 288}
    (b) feminine possessor {HP I Icelandic, 237}
        jafnvel [Alohomoragaldurinn   **[hennar**   Hermione]$_{POSSESSOR}$]
        even   [Alohomora_charm:DEF.ACC [3SG.F.GEN Hermione.GEN]]
        virkaði   ekki
        work:PST:3SG NEG
        '[. . .] not even when Hermione tried [**her** [Alohomora Charm]].' {HP I English, 300}

The proprial possessive construction is by no means a negligible quantity. In the Icelandic translation of the first volume of the Harry Potter series, there are altogether 228 tokens of adnominal possessive constructions which involve an ANTH-possessor. The inflectional genitive is used 136 times whereas we find 55 tokens for the proprial possessive with the remaining 37 tokens being reported for the prepositional constructions. Figure 17 shows the respective shares. The proprial possessive construction is responsible for slightly less than a quarter of all cases. It is attested more frequently than the prepositional constructions but it is ousted by the inflectional genitive with a ratio of 2.5-to-1. What needs to be explained is the competition between the three options since all three of them are licit with ANTH-possessors.

Thráinsson (2010: 95) classifies the proprial possessive construction as "default variant" in adnominal possession provided the possessor is an ANTH(-N). In HP I Icelandic, the default character of the proprial possessive construction does not come to the fore that clearly. Yet, it is certainly a major option when it comes to expressing adnominal possession. None of our secondary sources on Icelandic puts forward a hypothesis as to the factors which might determine the choice of the proprial possessive construction. We therefore check whether the possessee is in any way responsible for preferring one option over the other(s).

Figures 18–19 are helpful in this respect since they elucidate the differential preferences of constructions and possessive categories. The quantities result from the thorough analysis of HP I Icelandic, meaning: they are based on a single primary source which is a translation from an English original. A larger corpus of

4.3 Against parsimony: The pronominal possessive in Icelandic — 85

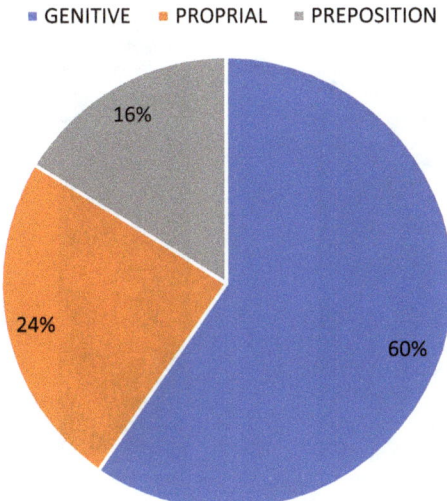

Figure 17: Competing adnominal possessive constructions in Icelandic.

genuine Icelandic texts of different genres might yield a more reliable and perhaps also very different picture. However, for the purpose of this section, the data from HP I Icelandic suffice to make our point. What we see immediately when we look at Figure 18 is that the proprial possessive construction is overwhelmingly used with a possessee whose referent is a concrete object (~ 73% of all attestations of the proprial possessive construction). As to the inflectional genitive, possessees with the feature [+concrete] are responsible for only 20% of all attestations. This class of possessees is even completely absent from the domain of the prepositional possessives. At the same time, body-part possession is highly important for the genitival construction with some 43% of all attestations and the prepositional constructions, almost 95% of which involve a possessed body part. In contrast, body-part possession is only marginally attested with the proprial possessive construction, namely exactly once which equals not even 2% of all attestations of the construction. The possessive PPs seem to be specialised the most because they are attested in combination with only two out of five possessee classes. The proprial possessive and the genitival construction are compatible with all classes of possessees albeit to different degrees.

This impression is corroborated by Figure 19 which surveys the shares the different constructions have within a given domain. The proprial possessive construction outnumbers its competitors in the domains of concrete possessees and the heterogeneous class of others. In these cases, the proprial possessive construction covers 60% of all tokens. In the remaining three domains, the share of the proprial possessive construction fails to exceed 17%. The genitival construction is the pre-

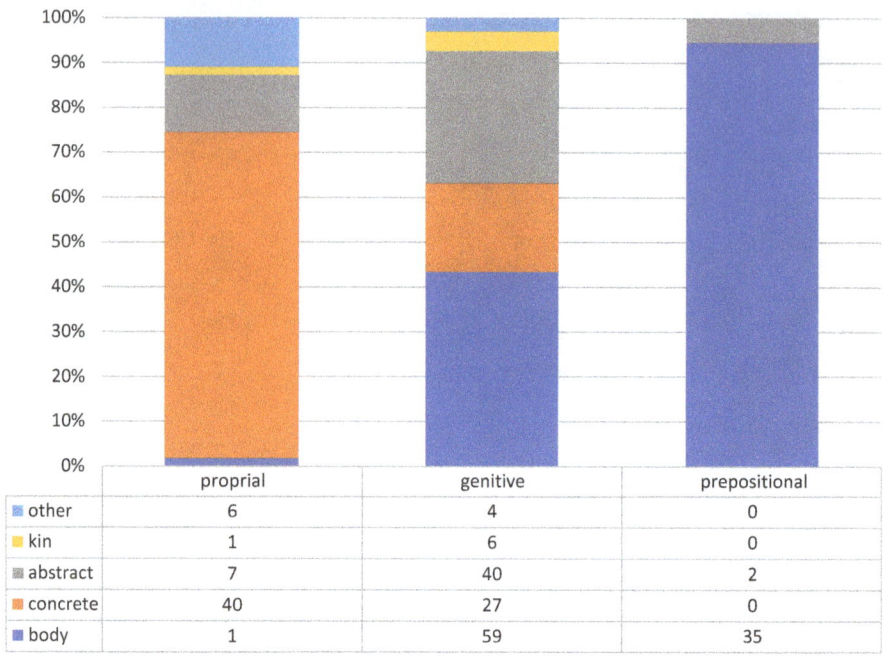

Figure 18: Shares of possessee classes over constructions.

ferred option with possessed body parts (~ 62%), abstract possessees (~ 81%), and kin (~ 86%). The prepositional possessives never form the majority solution. With 37%, their share is relatively high only in the case of body-part possession.

It is interesting to inspect cases of competition. Kress (1982: 229) argues that the genitival construction can be replaced with prepositional constructions, especially in the case of body-part possession or part-whole relations. The replacements are *í* 'in' and *á* 'on' (Kress 1982: 195 and 197). But what about the proprial possessive construction? To answer this question, we tick off the different possessee classes one by one by way of comparing examples of the proprial possessive construction with those of its competitors. We start from the bottom end, in a manner of speaking. In (48), body-part possession is illustrated – the domain where the proprial possessive construction is least frequently attested. We then proceed from right to left according to the order of the possessee classes in Figure 19 so that the domain where proprial possessives are particularly strong comes last: (49) involves kinship possession, (50) exemplifies the possession of abstract notions, (51) features the category of others, and (52) is dedicated to cases of possessed concrete objects.

4.3 Against parsimony: The pronominal possessive in Icelandic

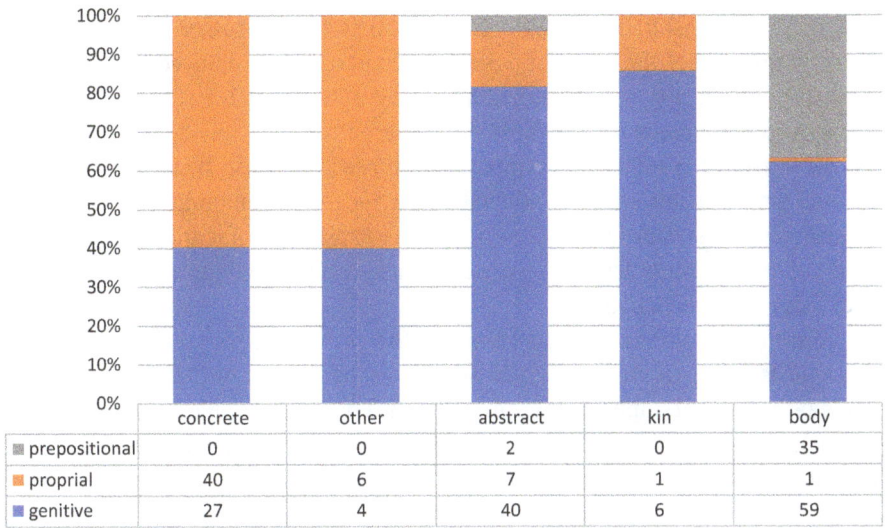

Figure 19: Constructions over possessee classes.

(48) Icelandic [Indo-European, Germanic] / body part
   (a) proprial {HP I Icelandic, 73}
      Þú   hefur   [augun   **[hennar   móður   þinnar]**$_{POSSESSOR}$]
      2SG   have:2SG   [eye:PL:DEF   [**3SG.F.GEN**   mother.GEN   **2SG.GEN**]]
      'You have [[your mother's] eyes].' {HP I English, 93}
   (b) genitival {HP I Icelandic, 211}
      Ævintýraþráin   skein   úr   [augum   [Ron-s]$_{POSSESSOR}$]
      thirst_of_adventure:DEF   shine.PST   out   [eye:DAT.PL   [Ron-**GEN**]]
      'The light of adventure was kindling again in [[Ron's] eyes] [. . .].' {HP I English, 267}
   (c) prepositional {HP I Icelandic, 109}
      Kennarinn   með   arnarnefið   leit   framhjá   Quirrell
      teacher:DEF   with   eagle_nose:DEF   look.PST   past   Quirrell
      professor   og   beint   í   [augun   **[á   Harry]**$_{POSSESSOR}$]
      professor   and   directly   in   [eye:PL.DEF   [**on**   Harry.DAT]]
      'The hook-nosed teacher looked past Quirrell's turban straight into [[Harry's] eyes] [. . .].' {HP I English, 138}

The co-existence of the genitival and the prepositional strategy with possessed body parts is explained by Thráinsson (2010: 94) as stylistically motivated with genitive constructions "tend[ing] to be more formal." This explanation probably covers (48b) and (48c) but leaves (48a) unaccounted for. The proprial possessive construction involves a kinship term as possessor, i.e., an ANTH-N possessor. Example (48a) is the only instance of the proprial possessive construction employed in the domain of kinship possession. It is taken from a paragraph in which Harry Potter's physical features are compared to those of his parents. The bodily properties he has inherited from his father are compared to those which can be attributed to his mother. Since the latter are mentioned only after those stemming from his father, contrastive focus comes to mind as a possible explanation for the use of the proprial possessive construction because this construction allows putting stress on the possessive pronoun *hennar* 'her'.

A similar argument can be used for the proprial possessive in (49a). The example is taken from a context in which the behaviour of different groups of people – parents of pupils – under the influence of the evil Voldemort is discussed. The behaviour of Malfoy's father contrasts with that of other people who remained loyal to the established system of the magicians' society.

(49) Icelandic [Indo-European, Germanic] / kin
 (a) proprial {HP I Icelandic, 96}
 Hann   segir   að   [pabbi   [hans   Malfoy-s]$_{POSSESSOR}$]
 3SG.M say:3SG that [father [3SG.M.GEN   Malfoy-GEN]
 hafi   ekki   þurft   neina   afsökun
 have:3SG.SUBJ NEG   need:PTCPL   NEG:ACC   excuse
 'He says [[Malfoy's] father] didn't need an excuse [. . .].' {HP I English, 121–122}
 (b) genitival {HP I Icelandic, 128}
 Þetta   er   asnalega   kúlan   sem   [amma
 DEM   be.3SG   stupid   globe:DEF   REL   [grandmother
 [Longbottom-s]$_{POSSESSOR}$]   sendi   honum.
 [Longbottom-GEN]   send:PST.3SG 3SG.M.DAT
 'It's that stupid thing [[Longbottom's] gran] sent him.' {HP I English, 162}

In (49b), no contrast is created since the magic globe is the only object on stage and its origin is known to all people present at the scene. There is thus no need for focus – and this is probably the reason why the proprial possessive construction is not used. Note that like (48a) in the previous case of body-part possession, (49a) is the only example of the proprial possessive construction being used to express kinship possession.

In contrast to body-part possession and kinship possession, the employment of the proprial possessive construction in (50a) is not as easily explained by way of invoking contrastive focus.

(50) Icelandic [Indo-European, Germanic] / abstract
 (a) proprial {HP I Icelandic, 210}
  um það bil   viku   fyrir   prófin   reyndi
  approximately   week:ACC   before   exam:DEF   test:PST.3SG
  óvænt   á   [nýja heitið   **[hans   Harry-s]**$_{POSSESSOR}$*]*
  unexpected   on   [new   promise:DEF [3SG.M.GEN   Harry-GEN]]
  um   að   hætta   afskiptaseminni.
  about   to   stop:INF   interference:DAT
  'Then, about a week before the exams were due to start, [[Harry's] new resolution] not to interfere in anything that didn't concern him was put to an unexpected test.'{HP I English, 266}
  lit. 'About a week before the exams [[Harry's] new promise] about stopping interference was put to the test unexpectedly.'
 (b) genitival {HP I Icelandic, 225}
  Hagrid   myndi   aldrei   bregðast   [trausti
  Hagrid   will:PST.3SG   never   fail:INF:RFL [trust:DAT
  [Dumbledore-s]$_{POSSESSOR}$*]*
  [Dumbledore-GEN]]
  'Hagrid would never betray Dumbledore.' {HP I English, 285}
  lit. 'Hagrid would never fail [[Dumbledore's] trust].'
 (c) prepositional {HP I Icelandic, 251}
  hann   [...]   heyrði   aðeins   [hryllileg   angistarópin   **[í**
  3SG.M   [...]   hear:PST.3SG   only   [terrible   shriek:DEF   [in
  Quirrell]$_{POSSESSOR}$*]*
  Quirrell.DAT]]
  '[...] he could only hear [[Quirrell's] terrible shrieks] [...].' {HP I English, 317}

However, in the wider context from which (50a) is taken, Hermione Granger, Harry's close friend, argues vehemently against getting involved again in matters that are not the friends' business. This is the same context from which (48b) stems. Example (50b) belongs to a succession of statements about the character Hagrid who is the focus of interest of Harry Potter's reasoning. Dumbledore's trust is compared neither to anyone else's trust nor to another physico-mental state of his. The choice of the prepositional possessive construction in (50c) is motivated as follows. Certain concepts like names, voices, bodily reactions, feelings, etc. are usually treated like body parts in Icelandic. Quirell's fearful cries are produced within his body and thus require the use of the preposition *í* 'in'.

The other category is largely an ad-hoc creation of ours. It contains mostly possessees which are in a social relation (other than kinship) with the possessor. The

COMM *klíka* 'gang' is attested five times in HP I Icelandic in combination with an ANTH-possessor which is always Dudley. In all five cases, the proprial possessive construction is chosen. The only other possessee type in the other category which triggers the proprial possessive construction is *fjölskylda* 'family'.

(51) Icelandic [Indo-European, Germanic] / other
    (a) proprial {HP I Icelandic, 29}
        enginn   vildi   komast   í   kast   við
        nobody want:PST.3SG come:INF:RFL in throw with
        *[klíkuna   [**hans**   Dudley-s]_{POSSESSOR}]*
        [gang:DEF.ACC [3SG.M.GEN Dudley-GEN]]
        '[...] nobody liked to disagree with [[Dudley's] gang].' {HP I English, 38}
    (b) genitival {HP I Icelandic, 24}
        Dursleyhjónin   höfðu   fengið   áminningarbréf
        Dursley_couple:DEF have:PST.3PL receive:PTCPL reprimand_letter
        frá   [skólastjóra   [Harry-s]_{POSSESSOR}]
        from [school_director:DAT [Harry-GEN]]
        'The Dursleys had received a very angry letter from [[Harry's] headmistress] [...].' {HP I English, 32}

The genitival construction occurs only four times but each time with a different possessee. This seems to indicate that the genitival construction is more common in terms of type frequency. The connection of the proprial possessive construction with contrastive focus is not straightforward, if at all. In HP I Icelandic, there is never any mention of a gang other than that headed by Dudley. The fear the gang inspires and the behaviour of its members are not compared to other gangs. The best we can do is guess that there is implicit or tacit comparison at work.

In the domain of alienable possession of concrete objects, the proprial possessive construction is at home, in a manner of speaking. Owing to its relatively high token and type frequency it also shows up in contexts where it is almost impossible to hold contrastive focus responsible for the choice of construction. This is different, however, with cases like those in (52).

(52) Icelandic [Indo-European, Germanic] / concrete
    (a) proprial {HP I Icelandic, 126}
        *[Kústurinn   [**hans**   Harry-s]_{POSSESSOR}] skaust   upp   í*
        [broom:DEF [3SG.M.GEN Harry-GEN]] shoot.PST up in
        höndina   á   honum
        hand:DEF.ACC on 3SG.M.DAT
        '[[Harry's] broom] jumped into his hand at once [...].' {HP I English, 160}

(b) genitival {HP I Icelandic, 126}[56]
[Kústur [Hermione Granger]*POSSESSOR*]   rúllaði   einfaldlega
[broom [Hermione.GEN Granger.GEN]]   roll:PST.3SG simply
frá   henni   og   [kústur [Neville-s]*POSSESSOR*] hreyfðist
from   3SG.F.DAT and [broom   [Neville-GEN]]   move:PST.3SG:RFL
ekki   hið   minnsta.
NEG   DEF.NT least
'[[Hermione Granger's] ___] had simply rolled over on the ground and [[Neville's] ___] hadn't moved at all.' {HP I English, 160}

The two sentences follow each other directly in one and the same paragraph in which it is described how the brooms react to the orders of their masters. As stated in (52a), Harry Potter's broom is exceptionally obedient whereas those of Hermione Granger and Neville Longbottom display strong resistance as shown in (52b). This is thus a bona fide case of contrast. Why the proprial possessive construction is made use of only once and why it is used in (52a) instead of being used in (52b) cannot be determined uncontroversially.

We refrain from elaborating further on this issue in this section because the interaction of the different semantic, pragmatic, syntactic, and stylistic factors is by far too intricate to be adequately described and evaluated in a study whose principal aims are outside the domain of possession. Even on the basis of our limited corpus, the assumption seems to be justified that SAG is firmly established in Icelandic which is a representative of Type IV. Since TOPOs are not featured as a separate topic in the literature, we assume that their behaviour resembles that of COMMs. It is clear, however, that the questions which remain need to be answered in a separate investigation. For the time being, we contend ourselves with postulating the split rule in Figure 20 which captures the gist of what has been said in the foregoing paragraphs. ANTHs and COMMs are not absolutely dissociated from each other since there is again a zone of overlap where both word-classes are allowed to function as possessors in the same construction type. Apart from this small zone of overlap, there are also other domains to which only one but not the other of the contenders is admitted.

A word of caution is in order in connection to the split rule in Figure 20. Given that for many native speakers of Icelandic, the proprial possessive construction is the default option when it comes to expressing adnominal possession with an ANTH-possessor (as claimed by Thráinsson 2010), the pragmatic criterion of focus

---

[56] The underlined empty spaces in the translation of (52b) indicate the position where the previously mentioned COMM *kústurinn* 'the broom' would be placed if equi-deletion had not applied.

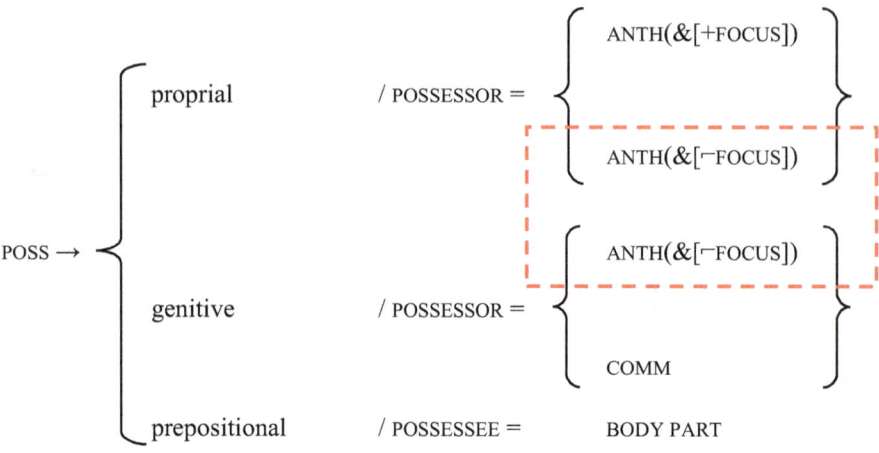

Figure 20: Split rule in Icelandic.

cannot be as important as Figure 20 suggests. It may be the case that focus is more important in the written register whereas it is largely irrelevant in spoken Icelandic. Irrespective of the possibility that Figure 20 reflects only one register of Icelandic, for two branches of the split rule, the crucial factor is the possessor (even if focus is removed from the list of criteria). In contrast, the choice of the prepositional strategy depends upon the semantic class of the possessee which must be a body-part term. At the moment, we cannot state with certainty whether the properties of the possessee overrule those of the possessor by default. This interesting problem will be taken up again in Section 4.5.

4.4 Neutralisation: The proprial gender in Romanian

Romanian [Indo-European, Romance] divides nouns into three genders. There are masculine, feminine, and neuter nouns (Loporcaro 2018: 92–109). Beyrer et al. (1987: 67) mention a fourth gender which they label "persönliches Genus", i.e., personal or proprial gender. This proprial gender is reserved for ANTHs. On what basis do Beyrer et al. (1987) assume the existence of the proprial gender? The raison d'être of the proprial gender transpires from the morphological and morphosyntactic behaviour of ANTHs which differs markedly from that of COMMs (Stolz et al. 2008: 396–404, Stolz et al. 2017a: 128–130). Before we inspect these differences a word of caution is in order. Strictly speaking, there is no proprial gender because ANTHs as controllers do not obey rules other than those which are valid for COMMs. If grammatical gender is defined as "classes of nouns reflected in the behaviour of associated words" (Hockett 1958: 231 quoted after Corbett 1991:1) then agreement

is crucial for the identification of gender distinctions (Corbett 1991: 4). However, ANTHs referring to male referents trigger the same agreement on their targets as masculine COMMs do as, e.g., with predicative adjectives in
(a) *băiatul* / *Vlad este blond* {boy:DEF.M} / {Vlad} {be.3SG} {blond.M} 'the boy / Vlad is blond'.

Similarly, ANTHs referring to female referents have the same agreement effects as feminine COMMs as, e.g., with predicative adjectives in
(b) *fata* / *Lisa este blond-ă* {girl:DEF.F} / {Lisa} {be.3SG} {blond-F} 'the girl / Lisa is blond'.

It is perhaps more appropriate to speak of a separate inflectional class to which only ANTHs are admitted. Below we will discuss a case of supposed disagreement which might give rise to the idea that female ANTHs have an agreement pattern which differs from COMMs and thus may claim the status of a distinct gender.

The special status of ANTHs in the morphological domain can be gathered from Table 11 where we contrast the paradigms of the COMMs *om* 'man' (masculine) and *casă* 'house' (feminine) with those of the ANTHs *Vlad* (male referent), *Letiția* and *Catrinel* (both female referents). For obvious reasons, neuter and plural forms are left out since they do not play any role in the ensuing discussion. Since ANTHs are inherently definite because of their monoreferentiality we also provide examples only of definite COMMs.

Table 11: Paradigms of COMMs and ANTHs in Romanian.

gender	case	COMM		ANTH			
M	NOM		om-ul		Vlad		
	ACC		om-ul	pe	Vlad		
	DAT		om-ului	lui	Vlad		
	GEN	(a)	om-ului	(a) lui	Vlad		
F	NOM		cas-a		Letiția		Catrinel
	ACC		cas-a	pe	Letiția	pe	Catrinel
	DAT		cas-ei		Letiți-ei	lui	Catrinel
	GEN	(a)	cas-ei	(a)	Letiți-ei	(a) lui	Catrinel
final segment			-a#			other	

A distinguishing feature of ANTHs as opposed to COMMs is that ANTHs are obligatorily accompanied by the preposition *pe* 'on' in the accusative. Accusative-marking by *pe* is also possible with animate (preferably human) COMMs but is subject to restrictions which are connected to definiteness. Caro Reina (2020b: 241–242) shows that *pe* is licit with indefinite COMMs whereas kinship terms like *mamă* 'mother' are made definite when combining with the accusative marker (cf. (53) – the grammaticality judgments being Caro Reina's).[57] There is thus reason to assume that Romanian too displays the extended category ANTH-N.

(53) Romanian [Indo-European, Romance] (Caro Reina 2020: 241–242)
    (a) indefinite human COMM[58]
        o   iubește   pe   fată   [*pe fat-a]
        3SG.ACC.F love:3SG on girl   [*on girl-DEF.F]
        'S/he loves a girl.'
    (b) definite kinship term
        o   iubește   pe   mam-a   [*pe mamă]
        3SG.ACC.F love:3SG on mother-DEF.F [*on mother]
        'S/he loves mom'. [O.T.]'
    (c) ANTH
        o   iubește   pe   Alina
        3SG.ACC.F love:3SG on Alina
        'S/he loves Alina'. [O.T.]'

Independent of the exact spellout of the rules which regulate the use of *pe* 'on' in case-marking function, these examples suffice to prove that there is SAG in Romanian since *pe* 'on' is mandatory with ANTHs but not with COMMs.

The differential distribution of *pe* 'on' is by no means the only ingredient of SAG in Romanian. What is more interesting for the topic of this section is the genitive. From Table 11, we already know that the genitive of COMMs is inflectional as additionally shown in (54).

---

57 Note that the distribution of *pe* is much more variegated than this statement makes one believe. Iordan and Robu (1978: 649–650), Mallinson (1986: 89), and Beyrer et al. (1987: 280–282) discuss a plethora of factors which determine the at times rather flexible use of the accusative marker. The examples in (53) should therefore be accepted only with a grain of salt.
58 In the original, the finite verb form is erroneously glossed as 2nd person singular. Likewise, the translation given is 'S/he loves the girl' with a definite direct object although *fată* 'girl' is formally indefinite.

(54) Romanian [Indo-European, Romance]
   (a) COMM – possessor = masculine {HP I Romanian, 86}
   noi   mergem   spre   [mijlocul   [tren-**ului**]$_{POSSESSOR}$]
   1PL   go:1PL   to   [centre:DEF.M   [train-**DEF.GEN.M**]]
   '[...] we are going down [the middle [**of the** train]] [...].'
   {HP I English, 109}
   (b) COMM – possessor = feminine {HP I Romanian, 87}
   Charlie   era   [căpitanul   [echip-**ei**   de
   Charlie   be.PST.3SG   [captain:DEF.M   [team-**DEF.GEN.F**   of
   Vâjthat]$_{POSSESSOR}$]
   Quidditch]]
   '[...] Charlie was [captain [**of** Quidditch]] [...].' {HP I English, 111}

The vast majority of ANTHs, however, do not inflect for case but use the prenominal form *lui* to encode the genitive (and dative) as shown in (55). For the time being, we gloss *lui* as X. There is genitive-dative syncretism in the case of syntactically light possessee NPs. If the possessee is represented by a heavy NP (with postnominal attributes, etc.), the possessor is additionally marked by *a* F.SG / *al* M/NT.SG / *ai* M.PL / *ale* F/NT.PL– labelled genitival or possessive article (Beyrer et al. 1987: 93–97) – which cannot be used to express the dative. In what follows, we do not further discuss the dative.

(55) Romanian [Indo-European, Romance]
   (a) POSSESSOR = ANTH with male referent {HP I Romanian, 128}
   globul   strălucea   în   soare   în   [mâinile
   globe:DEF.M   shine:PST.3SG   in   sun   in   [hand:DEF.F.PL
   [**lui**   Draco]$_{POSSSESSOR}$]
   [X   Draco]]
   'The Remembrall glittered in the sun as he held it up.' {HP I English, 162}
   lit. 'The globe shone in the sun in [[Draco**'s**] hands].'
   (b) POSSESSOR = ANTH with female referent {HP I Romanian, 43}
   să-l   opreşti   tu   pe   [fiul   [**lui**   Lily
   SUBORD-ACC.3SG.M   stop:2SG   2SG   on   [son:DEF.M   [X   Lily
   şi   al   lui   James]$_{POSSESSOR}$]
   and   GEN:M.SG   X   James]
   să   meargă   la   Hogwarts
   SUBORD   go:3SG.SUBJ   in   Hogwarts
   'Stop [[Lily an' James Potter**'s**] son] goin' ter Hogwarts!' {HP I English, 68}

The marker under inspection is used with ANTHs irrespective of the sex of their referents. This is remarkable insofar as *lui* is the genitive singular of *el* 'he', the pronoun of the 3rd person singular masculine, i.e., as a pronoun it would trigger the English translation as 'his'. The corresponding feminine pronoun of the 3rd person singular is *ea* 'she' with the genitive *ei* 'her' (Beyrer et al. 1987: 108). The pronominal use of *lui* and *ei* is illustrated in (56).

(56) Romanian [Indo-European, Romance]
    (a) masculine POSSESSOR {HP I Romanian, 15}
        [mătuşa   şi   unchiul   [lui]_{POSSESSOR}]   îi   vor
        [aunt:DEF.F   and   uncle:DEF.M   **[POSS.3SG.M]**]   3SG.DAT.M   FUT.3PL
        explica   totul
        explain.INF   all
        '[**His** [aunt and uncle]] will be able to explain everything to him [. . .].'
        {HP I English, 20}
    (b) feminine POSSESSOR {HP I Romanian, 19}
        [glasul   [ei]_{POSSESSOR}   ascuţit]   sfâşie
        [voice:DEF.M   **[POSS.3SG.F]**   sharp]   tear_apart:PST.3SG
        liniştea   casei
        silence:DEF.F   house:DEF.GEN.F
        '[. . .] it was [**her** [shrill voice]] which made the first noise of the day.'
        {HP I English, 25}
        lit. '[. . .] [**her** [sharp voice]] tore apart the quietness of the house.'

In contrast to *lui*, *ei* is never employed as prenominal genitive-dative marker of ANTHs. Beyrer et al. (1987: 67) argue that the proclitic *lui*, contrary to what normative grammar dictates, frequently precedes female ANTHs. Accordingly, cases like *lui Lily* 'Lily's' in (55b) would be understood as violating the agreement rules of Romanian because masculine and feminine gender are combined in one and the same construction. Is this a piece of evidence for the existence of a proprial gender in Romanian?

According to Table 11, female ANTHs are divided into two morphological subclasses. On the one hand, there are ANTHs like *Letiţia* which boast an inflectional genitive (in this case *Letiţiei* 'Letitia's') and on the other hand, female ANTHs like *Catrinel* behave like male ANTHs because they require the presence of *lui* and do not take genitival suffixes (thus **lui Catrinel** 'Catrinel's'). Iordan and Robu (1978: 370) describe this split as follows. Except under certain phonological conditions (to be disclosed below), the genitive of ANTHs is always marked by the prenominal *lui* irrespective of the sex of the ANTH's referent. Female ANTHs which end in the low vowel <a> /a/ either preceded by a velar stop <c> /k/, <g> /g/ or as final component of the diphthongs <oa> /wa/ and <ea> /ja/ take the inflectional genitive. This rule is repeated in Beyrer et al. (1987: 67). The idea of the phonologically constrained use of the inflectional genitive is taken up again by Gönczöl (2020: 45–46) who additionally claims that

> [i]f we have to put a foreign name or a female name which does not end in **-a** in the [genitive] or in the [dative], we add **lui** in front of the names, just like with male names. [original boldface] (Gönczöl 2020: 46)

In contrast, Cojocaru (2004: 38) assumes that all female ANTHs ending in *-a* take the inflectional genitive (independent of their origin) while all other ANTHs – irrespective of the sex of their referent – combine with the prenominal genitive marker *lui*. Peţan (2001: 234) states that in colloquial Romanian gender neutralisation also affects female ANTHs ending in *-a* so that the forms DAT *Letiţi-ei ~ lui Letiţia* and GEN *(a) Letiţi-ei ~ (a) lui Letiţia* coexist and Table 11 has to be adjusted accordingly.

In point of fact, the genitive of female ANTHs is subject to variation in contemporary Romanian. As can be seen from the examples in (57), double marking of the genitive is one of the options. The referent of the ANTH-possessor is the former Romanian female tennis star Simona Halep.

(57) Romanian [Indo-European, Romance]
    (a) inflectional genitive (ziar.com 4 mai 2023)
        [Primul antrenor [al Simon-ei Halep]_{POSSESSOR}]
        [first:DEF.M coach [GEN:M.SG Simona-GEN Halep]]
        lansează o teorie conspiraţionistă
        launch:PST.3SG INDEF theory conspiratory:F
        '[[Simona Halep's] first coach] launched a conspiratory theory [. . .].'
    (b) prenominal *lui* (monden 11 aprilie 2023)
        *El este [noul iubit [al **lui** Simona Halep]_{POSSESSOR}]*
        3SG.M be.3SG [new:DEF.M loved [GEN:M.SG X Simona Halep]]
        'He is [[Simona Halep's] new lover].'
    (c) prenominal *lui* + inflectional genitive (Adriana Bursch personal communication; original source SketchEngine roTenTen16)
        *[atitudinea plină de sportivitate [a **lui** Simon-ei Halep]_{POSSESSOR}]*
        [attitude:DEF.F full:F of sportiness [GEN.F X Simona-GEN Halep]]
        '[. . .] [[Simona Halep's] absolutely sporty attitude] [. . . .]'

Structurally, the three examples in (57) resemble each other. There is always a syntactically heavy possessee which requires the employment of the genitival marker *a* (+ agreement morphology reflecting the gender and number of the possessee). Simona Halep's first name ends in *-a*. It is therefore a candidate for taking the inflectional genitive. This is the case in (57a) – the word-form *Simonei* 'Simona's' is what we expect on the basis of Cojocaru's (2004) approach to the genitive of female ANTHS. From the point of view of Gönczöl (2020), the ability of the ANTH *Simona* to inflect for the genitive by suffixation would qualify the ANTH as autochthonous because foreign ANTHS are said to combine only with *lui*. This is what happens in (57b) where *lui* precedes the uninflected ANTH *Simona* yielding *lui Simona* 'Simona's'. This example does not only define the ANTH as foreign (because it fails to inflect for the genitive) but it also violates Cojocaru's (2004) rule according to which all female ANTHS in *-a* take the inflectional genitive. To top it all, example (57c) shows that combinations of prenominal *lui* and the inflectional genitive occasionally show up, too. We take this variation to indicate that the system is currently undergoing changes.

That the range of variation is wider than the literature normally admits can be gathered from (58).

(58) Romanian [Indo-European, Romance]
    (a) inflectional genitive {HP I Romanian, 140}
        dar   [cuvintele   [Hermion-ei]$_{POSSESSOR}$]   îl   puseseră
        but   [word:DEF.PL.F [Hermione-GEN]]   3SG.ACC.M put:PLUP:3PL
        pe   gânduri   pe   Harry
        on   thought:PL on   Harry
        'But Hermione had given Harry something to think about [. . .].'
        {HP I English, 177}
        lit. 'But [[Hermione's] words] put Harry on thoughts [. . .].'
    (b) prenominal *lui* {HP I Romanian, 242}
        *[buzele   [**lui**   Hermione]$_{POSSESSOR}$]   începură*
        [lip:DEF.PL.F [X   Hermione]]   start:PLUP:3PL
        să   tremure
        SUBORD   tremble:PLUP:SUBJ.3PL
        '[[Hermione's] lips] trembled [. . .].' {HP I English, 308}

The foreign female ANTH *Hermione* does not end in *-a*. Its foreign origin and the word-final segment render *Hermione* an absolutely unlikely candidate for taking the inflectional genitive. However, this expectation is not met in (58a) which involves the word-form *Hermionei* 'Hermione's'. In HP I Romanian, the genitive

Hermionei is attested six times whereas the use of the prenominal *lui* in combination with *Hermione* is a hapax legomenon, namely that given in (58b).

Cornilescu and Nicolae (2015: 104–106) characterise both *lui* and *-ei* as expletive article which are devoid of information about semantic and grammatical gender, its sole function being that of retaining syntactic definiteness and the ability to inflect for case. They argue that there are male ANTHs ending in *-a* which are also compatible with the inflectional genitive so that the gender distinction M ≠ F is blurred. In Old Romanian (Stan et al. 2016: 293–294), feminine ANTHs were accompanied by the proclitic *ei* = GEN-DAT of 3SG.F *ea* 'she' as opposed to *lui* = GEN-DAT of 3SG.M *el* 'he'. The gender distinction was functional with ANTHs, too. The proclitic *ei* became obsolete in the subsequent development (Dimitrescu et al. 1978: 235–237). One might therefore interpret the current variation as the prefinal stage of a longterm diachronic process whose terminus is the generalisation of *lui* as gender neutralising ANTH-marker. It is worth noting that this development seems to be unique to Romanian within the branch of Balkan Romance varieties (Caragiu Marioțeanu 1975).

In Figure 21, we provide a simplified summary of the intricate situation in Romanian. We have to gloss over most of the variation. There is the usual zone of overlap between ANTHs and COMMs. However, the bulk of the ANTHs is situated outside this zone. This is why we confirm that there is robust evidence of SAG in Romanian which we classify as a Type IV language.

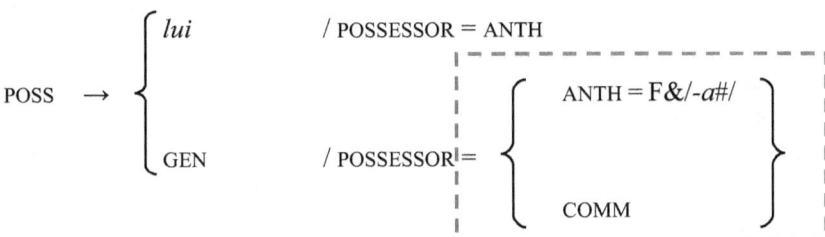

Figure 21: Split rule in Romanian.

A glance back at Table 11 reveals that SAG surfaces in a series of morphological mismatches. First of all, those ANTHs which take *lui* constitute an inflectional class of their own. Second, the exponent of the genitive in this inflectional class differs from that of COMMs. Where the latter comes as a suffix and thus occupies a slot on the right of the stem, *lui* is positioned to the left of the ANTH. We thus have the opposition of clitic vs affix or – if both marker-types are classified as clitics – that between proclitic and enclitic. Once again, ANTHs create structural heterogeneity but not in an arbitrary way.

4.5 Preliminary results: *Special Anthroponymic Grammar* in adnominal possession

In Sections 4.1–4.4, we have reviewed a number of SAG phenomena in the domain of possession. The many differences on the microlevel of individual languages notwithstanding, we recognise a pattern which is common to all of the languages discussed above. According to this pattern, ANTHs or a subgroup thereof are distinct from COMMs when it comes to expressing adnominal possession. Superficially, this is but a self-fulfilling prophecy since identifying cases of this kind was the main task of Section 4, in the first place. It is therefore hardly surprising that we have presented several such cases. However, what makes this necessarily incomplete collection of cases linguistically interesting is the cross-linguistic recurrence of the phenomena. We are not confronted with exceptions but with a sizable number of languages of different genealogy, structure, and location which share SAG properties which cannot be explained with reference to further structural traits they have in common, neither is it possible to treat the similarities as incidental. We argue that the correspondence between and across the above languages (and others tacitly passed over) is explicable with reference to general principles of possession.

Seiler (1983b: 26) has a short section in which COMM-possessors are compared to ANTH-possessors. He notes that in adnominal possession,

> proper nouns should not be considered here as representatives of the entire class of nouns. In fact, we note some differences in their respective behavior.

The behavioural differences in the morphosyntactic domain Seiler mentions are interpreted exclusively along the lines of the alienability correlation. The special status of ANTHs is not further discussed. To our minds, the focus on the opposition of alienable vs inalienable possession is by far too narrow – it might even be of minor relevance for the interpretation of the facts presented in Section 4 so far.

With reference to prior work by Taylor (1989), Heine (1997: 5 and 39) states that prototypically "[t]he possessor is a human being." The same opinion is expressed by Stassen (2009: 15) because the prototypical "possessor exerts control over the possessee." A similar view is held by Aikhenvald (2013: 11) who states that "[t]he possessor is often expressed with a personal pronoun or a proper noun." The prototypical possessor is human also for Lehmann (1998: 5–8), Fried (2009: 216), Mazzitelli (2015: 13), Schuster (2020: 35), to name only a few of the many scholars who have studied possession in-depth. It is not assuming too much when we conclude that the unanimity among experts of possession as to the prototypicality of human possessors has a negative side effect in the following sense. The properties of the possessor are taken for granted and backgrounded because the possessor is treated as the invariable component of the possessive situation. Much more attention is paid to the possessee since variation on this parameter frequently correlates with changes in the morphosyntactic expression of the possessive relation. We take issue with the onesided approach to possession by way of claiming that an inquiry into the properties of the possessor will yield new insights for linguistic possession research and, vice versa, the data from the domain of possession will shed new light on general principles of SAG in particular and SOG in general.

To prove that the common practice of ignoring what happens on the side of the possessor is detrimental to possession research, we start from the simplified schema in Figure 22. What we are going to argue is meant to hold for adnominal possession. Whether the situation is different in predicative possession must be clarified in a separate study.

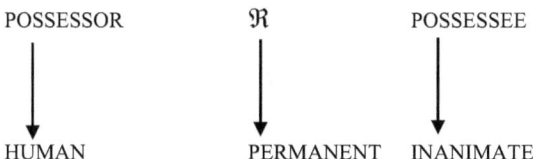

Figure 22: Prototypical possessive situation.

One way or another, this tripartite schema forms the basis of practically all linguistic work on possession referenced above (and dozens of related studies left unmentioned). For the sake of brevity, we skip definiteness as a frequently invoked factor. The three parameters POSSESSOR, RELATION, and POSSESSEE have prototypical realisations with HUMAN, PERMAMENT, and INANIMATE, respectively. Each of the parameters allows for variation in the sense that the prototypical realisation is replaceable with a different concept which may or may not be a subcategory of the prototype. Accordingly, under the rubric of POSSESSEE body-part terms, physico-mental states, etc. may take in the place of INANIMATE or an animate concept like kin can function as possessee. The structural correlates of the manipulation on the side of the possessee allow the researchers to detect, for instance, instances of the much-studied alienability-inalienability distinction (Heine 1997: 10–16). The possessive relation itself can be subject to variation – the prototypical PERMANENCE can be replaced with temporal or physical possession, etc. (Heine 1997: 38–39). As to the POSSESSOR, animacy is crucial for the prototype. Inanimate possessors are mentioned only for possessive situations which also involve an inanimate possessee. These are mostly part-whole relationships and thus examples of inalienable possession (Heine 1997: 35). Heine (1997: 40) depicts inanimate inalienable possession as far removed from the possessive prototype. On account of this reduced degree of prototypicality, inanimate possessors are not usually prominent in studies of possession. This means, in turn, that those possessive situations which are most commonly studied tend to involve animate (and this means human) possessors as a constant whose properties are presupposed and need not be looked into any deeper. The data discussed in the foregoing sections strongly suggest that this practice is a dead-end street.

If the possessor is prototypically human and (at least in examples in the linguistic literature) frequently represented by an ANTH this pattern is meaningful linguistically. Croft (2022: 69–70) presents the Extended Animacy Hierarchy based on referent salience in discourse. The categories involved in this hierarchy are said to "play a grammatical role in constructions" (Croft 2022: 69). ANTHs are mentioned repeatedly in this context but they somehow disappear from the discussion later on. Croft (2022: 70) presents two versions of the Extended Animacy Hierarchy which we reproduce here (with adjustments) as Figures 23–24.[59]

pronoun > PROP > COMM

Figure 23: Extended Animacy Hierarchy I.

1st/2nd person pronoun > 3rd person pronoun > (ANTH) > human COMM > animate COMM > inanimate COMM

Figure 24: Extended Animacy Hierarchy II.

The long version of the Extended Animacy Hierarchy in Figure 24 is only the more detailed version of the hierarchy in Figure 23. However, there is a striking difference between the two versions. In Figure 24, brackets are used for ANTHs. What the brackets are meant to signal is nowhere revealed. Since Croft (2022) does not refer to ANTHs in the remainder of his cross-linguistic study of morphosyntactic constructions, one might assume that the brackets tell the reader that no further information will be given as to ANTHs. What the brackets should not mean in any case is that ANTHs can be skipped because they are grammatically irrelevant in contrast to the other categories mentioned in the Extended Animacy Hierarchy.

Animacy is an important concept also in research on the grammar of PROPS (Van Langendonck 2007: 209, Nübling et al. 2015: 101–106, Helmbrecht et al. 2018, Thanner 2019). A high degree of animacy correlates with a high degree of prototypicality in terms of PROP-hood. ANTHs are at the top of the hierarchy. Except for pronouns, they also rank very high on Croft's Extended Animacy Hierarchy. ANTHs are thus doubly privileged, in a manner of speaking. For possession, this means

---

[59] Croft uses the symbol < which we replace with >. Instead of ANTH, Croft uses the term "human proper noun".

4.5 Preliminary results: *Special Anthroponymic Grammar* in adnominal possession — 103

that conceptually ANTHs are elementary and per default the first choice for the possessor role. In this way, ANTHs are principally entitled to structural properties which distinguish them from other, less prototypical possessors. Accordingly, to the logic of the Extended Animacy Hierarchy, we can expect ANTHs to behave differently from those nouny word-classes which are placed to their right in Figure 24.[60] There is thus an expectable if not absolutely predictable dividing line which separates ANTHs from human/animate/inanimate COMMs – and this is exactly what we have seen time and again in the foregoing sections. The behaviour of ANTHs in the domain of possession is fully in line with what the Extended Animacy Hierarchy implies. Our own Working Hypothesis (27a) is confirmed, too since there is ample evidence of rule splits along the lines of Figure 2.

The confirmation of our expectations is not the end of the story. It is also not enough to claim that the phenomena can be explained with reference to the Extended Animacy Hierarchy. The evidence for SAG in the domain of possession is a hard nut to crack if we look at them from the point of view of coding efficiency (Guzmán Naranjo and Becker 2021). Simplifying, the coding efficiency approach to language structures assumes that those components of a construction which are expected to occur are encoded as simply as possible with zero-marking serving as the potential optimum. In contrast, those components whose occurrence in a construction is not as predictable require overt and phonologically more substantial encoding (Haspelmath 2019). Given that ANTHs meet the criteria for the prototypical possessor, one would assume to find, in adnominal possession, juxtaposition of ANTH-possessor and possessee since the relation between the two participants is presupposed and thus needs no overt encoding. This juxtaposition strategy would ideally contrast with overt marking of some kind with COMM-possessors. However, only one of our case studies involves distinctive zero-marking in connection with ANTH-possessors. This is the case for Kryts where ANTH-possessors are zero-marked whereas COMM-possessors require the presence of the inflectional genitive as shown in Section 4.1. In the same section, Amele attests to the inverse distribution of overt marking and zero-marking because ANTH-possessors are accompanied by the possessive marker *na* while COMM-possessors remain unmarked. We do not deny that cross-linguistically many more cases like that illustrated with Kryts exist. Yet, the possibility that the pattern reflected in Amele might turn out not to be a

---

60 To the left of ANTH, Figure 24 only features pronouns. We do not go into the similarities and dissimilarities of pronouns and PROPs in this study. Nevertheless, it is worth mentioning that the Extended Animacy Hierarchy allows for the possibility that pronouns and ANTHs share properties which are not shared by COMMs. This possibility is especially interesting in the light of Anderson's (2007) hypothesis that PROPs are more like pronouns than they have things in common with nouns (cf. Section 2.2).

rarissimum (Cysouw and Wohlgemuth 2010) either cannot be excluded sweepingly as long as we have no reliably quantifiable cross-linguistic data. For the time being, we have to accept that contrary to the postulates of the coding-effiency approach, even the most prototypical and highly predictable participants of adnominal possessive constructions are not immune against being marked overtly while their less prototypical and less predictable counterparts receive no overt marking at all.

Coding efficiency is of course not the same as zero-marking. The relative shortness of the expression is the crucial criterion. Arguably, the *s*-genitive in several Germanic languages discussed in Section 4.2 fits this description because the ANTH-possessor takes the (monoconsonantal) genitive suffix whereas COMM-possessors form part of a PP which frequently requires the presence of the (in)definite article as well (provided the language possesses articles). COMM-possessors are thus part of the bulkier kind of construction. The *s*-genitive fulfils the criterion of coding efficiency. This probably also holds for the *sa*-genitive in Faroese. Superficially, the resumptive possessive of Low German is not as clear a case of the workings of efficient coding but might be explicable in this way on closer inspection. The proprial possessive in Icelandic (4.3), however, poses problems for coding efficiency since the additional possessive pronoun renders the construction more complex than those which are available for COMM-possessors. The pragmatic function of indicating contrastive focus can be achieved by prosodic means, i.e., the presence of the possessive pronoun is not the only structural option the language has to fulfil a given function. Similarly, the Romanian case described in Section 4.4 is difficult to explain satisfactorily by way of exclusively referring to coding efficiency. If the inflectional genitive is the most efficient strategy in the Germanic cases, why should it not be the same in Romanian? Given that this analogy is licit, it might be argued that Romanian COMM-possessors require the least complex construction.

The points we have raised in the foregoing paragraph do not prove coding efficiency wrong. Trying to do so would require a full-blown quantitative corpus-study for each of the languages mentioned above. This is beyond the scope of this study. What can be said nevertheless is that the morphosyntactic behaviour of ANTHs is a challenging domain for approaches like that of coding efficiency. More generally, SAG needs to be accounted for if one wants to model the network of factors which determine the shape of grammatical structure in human languages. Accordingly, the brackets around ANTH in Figure 24 should be cancelled out. In possession research, the fixation on the prototypical possessor should be given up in favour of a more differentiated viewpoint which allows us to distinguish between different kinds of human possessors. Moreover, scholars who take an interest in SOG – not only SAG – should seriously take functional aspects into consideration, meaning: the question should be posed how PROPs behave in morphosyntax when it comes to expressing certain notions or fulfilling certain functions. Only in this way will it be possible to generalise reliably over the grammar of PROPS language-independently.

The empirical part on SAG has featured a variety of cases, all of which can be subsumed under the heading of structural mismatches. These mismatches do not constitute a homogeneous class – and they are not confined to proper inflectional morphology. We will take up this issue again in the evaluation in Section 6. At this point, it is sufficient to state that zero-marking is by no means the only means to make ANTHS special as opposed to COMMS. Section 5 is dedicated to STG. Zero-marking is also a recurrent phenomenon in the domain of STG but by far not the only one as we will learn shortly.

5 Recurrent themes of *Special Toponymic Grammar*: Spatial relations

In this section we focus upon those properties which set TOPOs apart from COMMs and, if possible, also from other classes of PROPs when it comes to expressing spatial relations. That TOPOs tend to be zero-marked in constructions of spatial relations more often than COMMs has already been noted before, particularly in Stolz et al. (2014: 291). Haspelmath (2019: 315) sets forth that

> place names (=toponyms) are more usually in a locative role than other nouns, so when unexpectedly it is a common-noun argument that has the locative role, it has a greater need for special locative coding.

As will be demonstrated in the subsequent subsections, however, the special behaviour of TOPOs does not come to the fore only in the form of zero-marking. What we look out for is evidence of morphological patterns which are exclusive to TOPOs in general or sizable subclasses thereof without being shared necessarily by other nouny word-classes in their entirety. For the sake of brevity, we exclude questions of derivation and word-formation phenomena (such as the types of compounds) from our presentation. Fully lexicalised morphological idiosyncrasies of individual TOPOs are considered to be singularities to which we refer only unsystematically. We will start with languages in which STG manifests itself in properly grammatical aspects of morphology, viz. inflection (in the broadest sense of the term) and then turn over to languages in which syntactic properties come into play in addition to morphological aspects.

The principal aim is the identification of morphological (sub-)rules which apply specifically to (ideally all) TOPOs (of a given language). We do not only take stock of the evidence but also discuss the possibilities of systematising the cases under scrutiny in order to elaborate on how TOPOs are particularly prone to deviating from the morphological patterns of COMMs and sundry word-classes in the domain of spatial relations. The three basic spatial relations[61] considered in this study are
- Place, i.e., the location of an entity in space,
- Goal, i.e., the endpoint of the movement of an entity in space, and
- Source, i.e., the starting point of the movement of an entity in space.

---

[61] The three spatial relations correspond to the categories of general location *AT/TO/FROM* as put forward in the LDS questionnaire (Comrie and Smith 1977). The fourth category (*PAST*) is not taken into consideration in this study.

Furthermore, we follow Talmy (1985: 60–61), who defines a motion event as follows: "The basic motion event consists of one object (the 'Figure') moving or located with respect to another object (the reference-object or 'Ground')." A Ground can thus assume any of the roles (= Place, Goal, Source) in a spatial situation.

Figure 25 identifies the scenarios which are of interest to us in this section depending on which of the nouny word-classes fulfil the Ground role.

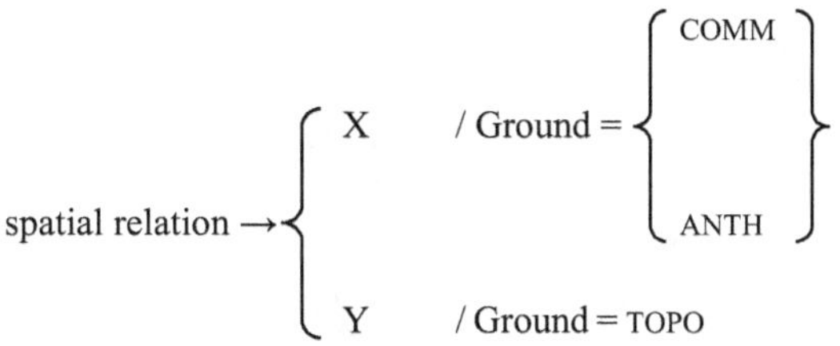

$$\text{spatial relation} \rightarrow \begin{cases} X & / \text{Ground} = \begin{cases} \text{COMM} \\ \text{ANTH} \end{cases} \\ Y & / \text{Ground} = \text{TOPO} \end{cases}$$

Figure 25: Schematic distinction of Grounds according to word-class.

In a given language, there are two (or more) constructions available for encoding spatial relations. Construction X is chosen if the Ground is a COMM (or ANTH) whereas construction Y is reserved for Grounds that belong to the class of TOPOS. Despite the oversimplified representation, Figure 25 illustrates the kind of cases that are of interest in the subsequent subsections, viz. cases in which TOPOS (or a group of TOPOS) behave differently from COMMS (and ANTHS) when serving as Ground in constructions encoding spatial relations. To get a feel for the cases under scrutiny, we first sketch a telling example from Lezgian (= Section 5.1) which, in turn, serves as the backdrop for parallel cases from a number of Uralic languages (= Section 5.2). In contrast to the Lezgian case, the Uralic evidence will be discussed in some detail. After evaluating the Uralic cases, we will turn our attention to further languages (such as the isolate Basque in Section 5.3) whose morphological system displays peculiarities in connection with TOPOS. In Section 5.4, evidence of the special grammar of TOPOS from Ancient Greek and Latin is presented. We also discuss data from Swahili in Section 5.5 which forms a kind of bridge that connects the evaluation of the phenomenology in the realm of morphology to what we encounter in the domain of morphosyntax to be addressed in follow-up studies.

5.1 Glimpses of Lezgian case-inflection

A good starting-point for the subsequent cross-linguistic account of the distinctive morphological behaviour of TOPOS is provided by Haspelmath (1993: 99) in his reference grammar of the Northeast Caucasian [Nakh-Daghestanian, Lezgic] language Lezgian where he observes that

> [t]he Superessive is also used with a number of names of Lezgian villages (other Lezgian villages and all non-Lezgian places take the Inessive) [...].

Accordingly, there is a contrast between, on the one hand, *Q'asumxür.e-l* 'in Q'asumxür (Kasumkent)', which expresses the static location at a place via the superessive *-l*, and, on the other hand, *Xiv.d-a* 'in Xiv' and *Leningrad-a* 'in Leningrad', both of which host the inessive suffix *-a* which fulfils the same function, namely that of indicating static location. There is thus morphological variation across TOPOs because one and the same function (namely that of Place) is expressed by different means. A minority of typically local TOPOs requires the employment of the superessive case whereas the vast majority of TOPOs are inflected for the inessive in the same function.

Moreover, this variation within the class of TOPOs also marks a split which separates a subclass of TOPOs from the bulk of the COMMs in Lezgian. The paradigm of Lezgian COMMs comprises eighteen cases, fourteen of which can be considered to be spatial. A COMM like *sew* 'bear' yields an inessive *sew-re* 'in the bear' which is distinct from the superessive *sew-re-l* 'on the bear' (Haspelmath 1993: 74). In the absence of any explicit statement as to the number of distinct cases of TOPOs, we have to make do with formulating a conjecture. In contrast to COMMs, TOPOs do not seem to inflect for the same number of case categories. We hypothesise that a TOPO has either an inessive or a superessive – but never both cases. If a TOPO is involved in a situation of spatial superiority, postpositions are employed in lieu of the inflectional cases of the so-called super localisation (Haspelmath 1993: 78–79) as shown, for instance, by the postpositional phrase [[*Berlin.di-n*$_{genitive}$]$_N$ **winel**$_{postposition}$]$_{PostP}$ '**above** Berlin' (Haspelmath 1993: 214).[62] This means that TOPOs are one case short in contrast to COMMs (cf. Table 12).

Table 12: Selected spatial cases of Lezgian COMMs and TOPOs.

case	COMM	TOPOS	
	sew 'bear'	*Tiv*	*Q'asumxür*
inessive	sew-re	Tiv.d-a	---
superessive	sew-re-l	---	Q'asumxür.e-l

---

62 Haspelmath (1993: 74) adds the remark that "the specific local relations that were originally expressed by the local cases are now generally expressed by postpositions, while the local cases mainly express more abstract senses." This proviso does not seem to hold for cases of general location such as the superessive and the inessive, though, which are described as fulfilling first of all the spatial functions of *ON ~ ONTO* and *IN ~ INTO*, respectively (Haspelmath 1993: 98 and 102).

Where COMMs formally distinguish two cases from each other, TOPOs have only one of the two possible case categories. Thus, two major patterns arise – one which holds for COMMs and the other which is representative of TOPOs. In the latter class, two subclasses exist which differ from each other as to which of the inflectional cases is used to encode the spatial relation of Place (= general location *AT ~ IN ~ ON*). This difference yields two inflectional classes. Class A comprises those TOPOs which take the inessive inflection as opposed to the members of Class B which are inflected formally for the superessive. The differences between the two inflectional classes on the formal level are not correlated to differences on the functional or semantic level since in both cases the intended meaning is location *AT*. Thus, Table 12 can be reorganised on purely functional grounds to yield a different picture (cf. Table 13).

Table 13: Selected spatial functions and their expressions with COMMs and TOPOs in Lezgian.

spatial relation	COMM	TOPOS	
	sew 'bear'	*Tiv*	*Q'asumxür*
AT/IN	*sew-re*	*Tiv.d-a*	*Q'asumxür.e-l*
ON/ABOVE	*sew-re-l*	*Tiv.d-in winel*	*Q'asumüxer-in winel*

The paradigm of the TOPOs is not properly defective (Baerman and Corbett 2010: 4–11) since the putatively empty cells can be filled regularly with postpositional phrases[63] so that we are dealing with a case of paradigmatic **periphrasis** (Haspelmath and Sims 2010: 183). It could be argued, however, that the use of the superessive case for the purpose of expressing general location *AT* (which is normally expressed by the inessive case) illustrates an instance of lexical-paradigmatic **deponency** (Baerman 2007: 10–13). These and sundry morphological mismatches will be referred to time and again in the subsequent paragraphs.

The Lezgian appetiser is indicative of the special morphological behaviour of TOPOs in contrast to the patterns which are reflected by the paradigms of the vast majority of the COMMs of the same language. Superficially, the structural differences of TOPOs and other classes of nouns may not be particularly striking and thus tend to escape being noticed. However, they are recurrent and systematic never-

---

[63] According to Baerman and Corbett (2010: 5) "[t]he most obvious example of a defective paradigm is one in which some form is missing, i.e., where the gap can be described as the absence of some morphological entity, and we can base our morphological typology on the typology of components."

theless so that two distinct rules are required to describe the Lezgian situation (cf. Figures 26–27).[64]

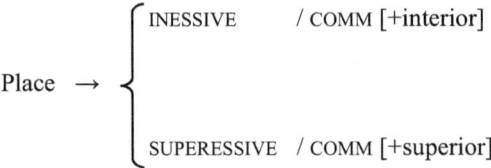

Figure 26: Choice of spatial cases with Lezgian COMMS

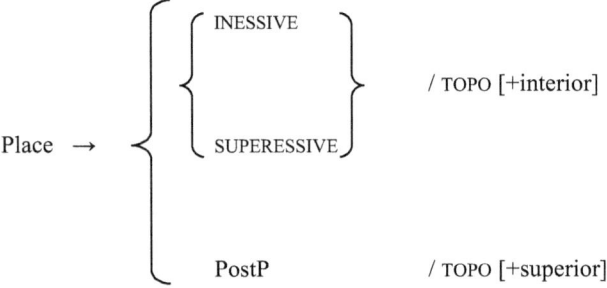

Figure 27: Choice of spatial cases with Lezgian TOPOS.

If we try to fuse the two rules into one, the product is relatively complex since several options have to be accounted for in one and the same formula. This means that TOPOS cannot be subsumed under the class of COMMS without increasing the complexity of the rules which describe the morphological behaviour of the members of these classes. Alternatively, two parallel but nevertheless distinct rules can be postulated (as shown above). The individual rules are less complex than the integrated super-rule would be. However, the transparency thus gained makes it necessary to assume a higher number of rules to be operative in the language. This dilemma is typical of those cases in which the morphologies of TOPOS and COMMS – however slightly – differ. The subsequent examples from Uralic support this idea empirically.

---

64 For other cases which were associated originally with the spatial relation Place – like the adessive, postessive and the subessive – Haspelmath (1993: 90, 92, and 95) – as mentioned in footnote 62 – states that the erstwhile "locative" function has been taken over largely by postpositions whereas the more abstract (i.e., non-spatial) functions of these cases have been retained. Since they are no longer used pirmarily with a spatial meaning, we exclude them from our account.

5.2 Uralic parallels

The above scenario is not unique to Lezgian. Similar situations are reflected in the grammars of the Uralic languages Hungarian, Finnish, and Estonian. Their relatively close genetic relationship notwithstanding, the three languages do not yield absolutely identical patterns. What is common to the three of them and Lezgian is, on the one hand, the different use TOPOS and COMMS make of their rich case systems and, on the other hand, the distinction of several sub-classes of TOPOS on the basis of their morphological behaviour. We look at each of the languages separately starting with Hungarian, the Uralic language with the largest number of native speakers, and continue with Finnish and Estonian in the order of the decreasing size of the relative speech-communities. In each of the sub-sections, we look first at the names of settlements such as cities, towns, and villages. Thereafter the names of more extended territories (as, e.g., counties, districts, and countries) and occasionally microtoponyms (as, e.g., streetnames) will be touched upon, too.

5.2.1 Hungarian

The exact number of cases of Hungarian is a matter of dispute. To cut a potentially never-ending story short, we follow Tompa (1972: 109–124) who assumes a system of maximally twenty-seven cases, several of which are presented either as hardly productive if at all or as limited to a reduced set of nouns or other word-classes. If we discount the unproductive and less commonly employed cases, there remains a paradigm of twenty-one cases which exceeds the already sizable system of Lezgian by three cases. Ten of the Hungarian case categories can be considered spatial, namely
– the three cases of the interior set (= inessive, illative, elative),
– the three cases of the superior set (= superessive, sublative, delative),
– the three cases of the exterior set (= adessive, allative, ablative), and
– the isolated terminative case.

Hungarian COMMS and PROPS[65] normally inflect for all of the above cases as, e.g., the COMM *hajó* 'ship' (only the singular inflections are given):

---

65 Cf. for instance the first name *Thomas* which is inflected for the delative in {Róz 151} *A kis kamra fiókjában volt egy fénykép* **Thomas-ról**. . . 'In the chest of drawers of the junk room, there was a photograph **of Thomas**. . .' Note that, in this example, the delative has no properly spatial meaning. Other examples are allative *Hagridhoz* 'to Hagrid' {HP I Hungarian, 245}, sublative *Harryre* 'onto Harry' {HP I Hungarian, 168}, etc.

- *hajó* 'ship' → inessive *hajó-ban* '**in** the[66] ship', illative *hajó-ba* '**into** the ship', elative *hajó-ból* '**out of** the ship', superessive *hajó-n* '**on** the ship', sublative *hajó-ra* '**onto** the ship', delative *hajó-ról* '**down from** the ship', adessive *hajó-nál* '**at/near** the ship', allative *hajó-hoz* '**to** the ship', ablative *hajó-tól* '**from** the ship', terminative *hajó-ig* '**as far as** the ship' (Tompa 1972: 123–124).

TOPOs have the same inflectional capacity as the average COMM. However, the domain of the case categories differs markedly from that of the same categories with COMMs. Superficially, one might get the idea that the paradigms of TOPOs are defective since in contrast to COMMs, TOPOs employ either only the cases of the interior set or only those of the superior set but never both sets for the purpose of expressing general location (but cf. below).

As to the morphology of geonyms referring to settlements Tompa (1972: 100) explains that

> die Siedlungsnamen als Ortsbestimmungen in ein und denselben Funktionen verschiedene Relationssuffixe erhalten [. . .]. Einige können auf die Frage wo? auch das heute veraltete Suffix -*t*, -*tt* annehmen [. . .]. Aber selbst diese Städtenamen werden in der Umgangssprache eher mit dem Inessivsuffix -*ben/-ban* [. . .] oder aber mit dem Superessivsuffix -*n*, -*en/-ön/-on* [. . .] gebraucht.[67]

There is thus a pattern of variation which is reminiscent of the situation as described for Lezgian above. This similarity does not mean that both cases are identical. In point of fact, where Lezgian restricts the variation to the spatial relation of Place, Hungarian shows that all three of the spatial relations of Place, Goal, and Source are affected by variation – and to different degrees at that, since the variation with Place comprises more options than that of Goal and Source. There are two major classes of TOPOs,[68] namely those which exclusively take cases of the superior set as, e.g.,
- *Budapest* → superessive *Budapest-en* '**in** Budapest', sublative *Budapest-re* '**to** Budapest', delative *Budapest-ről* '**from** Budapest'

---

[66] We add the definite article in the English translation for purely stylistic reasons. Accordingly, the Hungarian examples are treated throughout as if they contained the indeclinable definite article *a* 'the' as in *a hajóban* 'in the ship'.

[67] Our translation: "as spatial adverbial, the names of settlements host different relational suffixes for one and the same function [. . .]. In answer to the question *where?*, some may take the nowadays obsolete suffix -*t*, -*tt* [. . .]. However, in the colloquial language, even these names of cities are used more frequently with the inessive suffix -*ben/-ban* [. . .] or the the superessive suffix -*n*, -*en/-ön/-on* [. . .]."

[68] All Hungarian examples in this paragraph are taken from Tompa (1972: 100).

and those which employ the cases of the interior set instead as, e.g.,
- *Bécs* 'Vienna' → inessive *Bécs-ben* '**in** Vienna', illative *Bécs-be* '**to** Vienna', elative *Bécs-ből* '**from** Vienna'.[69]

Within both of these classes,[70] two sub-classes can be distinguished because apart from the majority option which takes the regular cases of the interior or of the superior set, there is also a minority of TOPOS which at least optionally makes use of the unproductive locative suffix *–t(t)* – besides the colloquially preferred inessive or superessive suffix as, e.g.,
- *Pécs* → locative *Pécs-ett* ~ superessive *Pécs-en* '**in** Pécs' and *Győr* → locative *Győr-ött* ~ inessive *Győr-ben* '**in** Győr'.

In (59), we provide sentential examples of the use of the superessive (= (59a)) and the inessive (= (59b)) as a means of encoding the spatial relation of Place with TOPOS.

(59) Hungarian [Uralic, Hungaric] – encoding of Place on TOPOS
    (a) superessive {Róz 50}
        Mire   **[Pest-en]**   leszek   megint, te   megnősz
        what:SUBLV **[Pest-SUPESS]** be.FUT:1SG again   you   grow_up:2SG
        'For as long as I am **[in Pest]** again, you will grow up [. . .].'
    (b) inessive {HP I Hungarian, 68}
        Harry   még   soha   nem   volt   **[London-ban]**.
        Harry   still   never   NEG   be.PRET **[London-INESS]**
        'Harry had not been **[in London]** yet.'

In (60) and (61), we illustrate the variation of the residual locative with the superessive (= (60)) and the inessive (= (61)) in the same function. It has to be borne in mind that the residual locative is possible neither on ANTHs nor on COMMs.

---

69 It must be mentioned here that the cases of the interior set and the superior set seem to be in free variation with COMMs at least in some contexts as in *A kalap a fejem-en* ~ *A kalap a fejem-ben* 'The hat is **on** my head.' (Tompa 1972: 116). However, this supposedly free interchangeability does not seem to be general.

70 As in the Lezgian example, the exterior set of cases in Hungarian is not affected by morphological processes which would single out TOPOS as different from other classes of nouns. This is the reason why we do not discuss the adessive, allative, and ablative in this section.

(60) Hungarian [Uralic, Hungaric] – locative vs superessive
   (a) superessive [Hungarian National Corpus[71]]
   Úgy   hírlett,   **[Pécs-en]**   már   bent   vannak   az
   so   announce:PRET   **[Pécs-SUPESS]** already   inside   be:3PL   DET
   oroszok.
   Russian:PL
   'So it was broadcasted, the Russian are already **[in Pécs]**.'
   (b) locative (Mikesy 1978: 396)
   Holnap   reggel   nyolckor   **[Pécs-ett]**   kell   lennem
   tomorrow   morning   eight:TEMP   **[Pécs-LOC]**   must   be.OPT:1SG.DEF
   'Tomorrow morning I have to be **[in Pécs]** at 8 o'clock [. . .].'

(61) Hungarian [Uralic, Hungaric] – locative vs inessive
   (a) inessive [Hungarian National Corpus[72]]
   hogy   **[Győr-ben]**   ölik   a   prostitualtakat
   that   **[Győr-INESS]** kill:3SG   DET   prostitute:PL:ACC
   '[. . .] that he kills the prostitutes **[in Győr]** [. . .].'
   (b) locative [Hungarian National Corpus[73]]
   Ő   már   **[Győr-ött]**   szolgálta   tovább   a   császárt.
   he   already   **[Győr-LOC]**   serve:PRET.DEF   further   DET   emperor:ACC
   'Already **[in Győr]** he served the emperor further.'

An informal check of the electronic Hungarian National Corpus has yielded 1,696 hits for the superessive *Pécsen* as opposed to 9.277 hits for the locative *Pécsett*, i.e., the latter accounts for 85% of all cases of the static Place relation expressed on the TOPO *Pécs*. As to the TOPO *Győr*, there are 2,217 tokens of the locative *Győrött* and 4.844 tokens of the inessive *Győrben*. In this case, the share of the inessive equals 69% of all instances of the spatial relation of Place being expressed on the TOPO *Győr*. Going by these figures, one cannot generalise sweepingly about the vitality of the residual locative since individual TOPOs seem to have different preferences as to which case is employed to encode Place.

Table 14 captures the above morphological differences in a systematic way. The data in this table reflect exclusively the situation which emerges in connection to the expression of the categories of general location. Grey shading highlights empty cells. Boldface marks the obsolete locative.

---

71 http://corpus.nytud.hu/mnsz/index_eng.html   (lit_hu_dia_Bertha_Bulcsu_Irok_muhelyeben_1973.clean).
72 http://corpus.nytud.hu/mnsz/index_eng.html (off_hu_misc_fovkozgy).
73 http://corpus.nytud.hu/mnsz/index_eng.html (lit_gr_tanczos_radvany.s1.clean).

Table 14: Spatial cases of the interior and superior set of Hungarian TOPOS and COMMS – expression of general location.

set	case	TOPOS				COMM
		Szeged	*Pécs*	*Debrecen*	*Győr*	*kör* 'circle'
interior	inessive	---	---	*Debrecen-ben*	*Győr-ött* ~ *Győr-ben*	*kör-ben*
	illative	---	---	*Debrecen-be*	*Győr-be*	*kör-be*
	elative	---	---	*Debrecen-ből*	*Győr-ből*	*kör-ből*
superior	superessive	*Szeged-en*	**Pécs-ett** ~ *Pécs-en*	---	---	*kör-ön*
	sublative	*Szeged-re*	*Pécs-re*	---	---	*kör-re*
	delative	*Szeged-ről*	*Pécs-ről*	---	---	*kör-ről*

COMMs allow for six distinctions in the realm of the interior and superior sets. TOPOs, however, are down to three distinctions which are made either within the interior set or within the superior set. As in the previous case of Lezgian, the paradigms of the TOPOs give the impression of being defective. On closer inspection, however, this impression turns out to be largely unjustified since the situation can be shown to be considerably more intricate.

In contrast to the situation in the Northeast Caucasian language, the Hungarian facts need to be scrutinised more closely. The three cases which refer to superior location in Hungarian tend to imply physical contact of Figure and Ground in the spatial situation.[74] For superior location without contact postpositional phrases with the postpositions *felett* ~ *fölött* 'above' (= Place), *fölé*[75] 'over' (= Goal), *felül* ~ *fölül* 'from over' (= Source) (Mikesy 1978: 135) are employed in combination with all kinds of nouns – including TOPOS. Like many other Hungarian postpositions (Bárczi 2001: 83–86), they have been grammaticalised from erstwhile relational nouns and host vestiges of an earlier triplet of spatial cases, viz. the already familiar locative in *-tt*, the lative in *-é*, and the ablative in *-ül*. The periphrastic construction that is used to express superior static location is exemplified in (62).

---

[74] This statement has to be taken with a grain of salt because Tompa (1972: 116) enumerates several cases of nouns for which the superessive is ambiguous as to the distinction of [+contact] vs [-contact] such as *kép* 'painting' → superessive *kép-en* 'on / over / above the painting'. In connection to other nouns, however, the absence of contact of Figure and Ground seems to be ruled out (as e.g. *borjú* 'calf' → superessive *borjú-n* 'on the calf').

[75] The cognate postposition *felé* 'in the vicinity of' has undergone a different development semantically so that it can no longer be used with reference to superiority (e.g. ***Szeged felé megálltunk egy vendéglőben.*** 'In the vicinity of Szeged, we stopped at an inn.' (Hessky et al. 2005: 382)).

(62) Hungarian [Uralic, Hungaric] {HP I Hungarian, 19} – Place [+superior&-contact]
 (a) TOPO
 [Bristol fölött] röpültünk, amikor elszunyókált.
 [Bristol above] fly:PRET:1PL when doze_off:PRET.3SG
 'We flew **[over Bristol]** when he fell asleep.'
 (b) COMM
 Nekem például *[a bal térdem fölött]* van
 DAT:1PL for_example [DET left knee:POR.1SG above] be
 egy heg
 a scar
 'I have for instance a scar **[above my left knee]** [...].'

This means that the paradigm of Hungarian TOPOs does not attest to periphrasis. Nevertheless, it cannot be considered to be genuinely defective since TOPOs display the full array of case distinctions if these are required by the head of a given construction. There is a sizable number of postpositions, for instance, which govern the superessive or sublative on their complement (cf. Figure 28) whereas verbs like *beszél* 'to talk about', *gondol* 'to think about', and *elfeledkezik* 'to forget something' etc. take an internal argument in the sublative or delative.

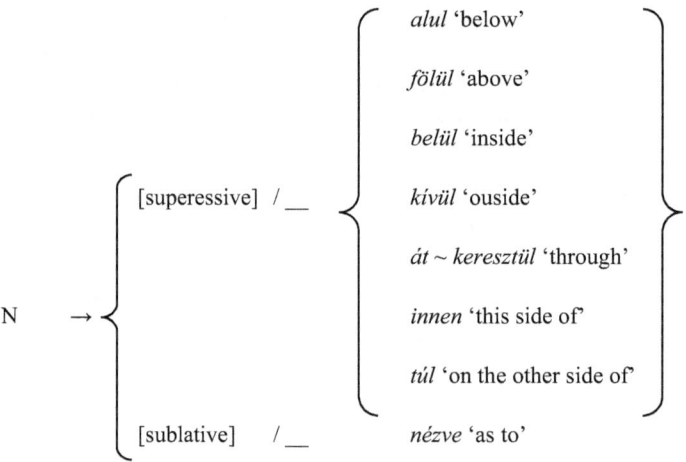

Figure 28: Postpositions governing cases of the superior set.

Accordingly, TOPOs which allow only the use of the spatial cases of the interior set in the domain of general location inflect for those of the superior set if this is required by the governing head. In (63), we provide sentential examples which involve names of countries which – with one exception to be discussed below – take the suffixes of the interior set to encode general location (as in (63a)). However, in postpositional phrases with one of the abovementioned heads, the very same names of countries behave like COMMs insofar as they inflect for cases of the superior set.

(63) Hungarian [Uralic, Hungaric] – names of countries
    (a) interior set {Táv 169}
        és évekig éltem [Olaszország-ban].
        and year:PL:TERM live:1SG [Italy-INESS]
        '…and I have been living for years **[in Italy]**.'
    (b) superior set {Táv 177}
        Hosszú volt az út és fárasztó
        long be:PRET DEF journey and tiring
        [[Törökország-on], [Bulgáriá-n], [Jugoszláviá-n],
        [[Turkey-SUPESS] [Bulgaria-SUPESS] [Yugoslavia-SUPESS]
        [Olaszország-on] és [Franciaország-on] át]$_{PostP}$
        [Italy-SUPESS] and [France-SUPESS] through]$_{PostP}$
        'The journey [through **[Turkey, Bulgaria, Yugoslavia, Italy and France]**] was long and tiring.'

In (63b), the postposition *át* 'through' triggers the superessive case on the preceding names of countries, each of which tolerates only cases of the interior set when general location is to be encoded. Similarly, the TOPO *Bagdad* 'Baghdad' is compatible exclusively with the cases of the interior set if spatial relations of the general-location kind are to be expressed (as in (64a)). The monopoly of the interior cases is lifted, however, if the TOPO is the complement of a verb whose case-frame requires its internal argument to be inflected for a case of the superior set (as in (64b)).

(64) Hungarian [Uralic, Hungaric]
    (a) interior set {Táv 6}
        Joan Scudamore [Bagdad-ból] tért vissza Londonba
        Joan Scudamore [Bagdad-ELA] turn:PRET back London:ILL
        'Joan Scudamore returned **[from Bagdad]** to London…'
    (b) superior set {Táv 179}
        Averil Barbaráról és [Bagdad-ról] kérdezősködött.
        Averil Barbara:DEL and [Baghdad-DEL] ask_vaguely:PRET.3SG
        'Averil asked vaguely**[about]** Barbara and **[Baghdad]**.'

Example (64b) also shows that TOPOs and ANTHs behave the same – and this means that they follow the pattern of COMMs. Thus, under the conditions of postpositional and verbal case government, the morphological differences are neutralised. It is therefore not entirely correct to assume defective paradigms for all of the Hungarian TOPOs.

Defectiveness can be assumed – if at all – only for those TOPOs which combine with cases of the superior set to express general location. Their supposed defectiveness is causally connected to the absence of patterns of government which involve the cases of the interior set, i.e., there are no verbs or postpositions which require their complement to be in the inessive, illative, or elative. Given that no head governs any case of the interior set, what appears to be an instance of morphological defectiveness turns out to be something else, namely the lack of syntactic contexts in which the putatively missing cases would be called for. In lieu of giving the label of proper defectiveness to the Hungarian constellation of facts, we therefore prefer to classify this case as an example of **pseudo-defectiveness**. Table 15 is meant to show that pseudo-defectiveness affects only TOPOs of Type A, i.e., of the class of TOPOs which encode general location by the suffixes of the superior set.

Table 15: Two different TOPO types and pseudo-defectiveness / Hungarian.

TOPO type	set	function	
		general location	other
A	superior	yes	yes
	interior	no	no
B	superior	no	yes
	interior	yes	no

Pseudo-defectiveness applies if the supposed gap in the paradigm is not an independent property of the morphological paradigm as such but the effect of the absence of an external trigger which would enforce the use of the missing caseforms. If there were verbs and postpositions in Hungarian which govern cases of the interior set, the TOPOs of Type A would no longer give the impression of being defective morphologically.

The Hungarian paradigms of TOPOs also give evidence of **overabundance**[76] because in two of the sub-classes there is allomorphy in the sense that the speakers have the choice of two synonymous options if they want to express the spatial relation of Place, namely either by the regular suffixes of the inessive *-bVn* and superessive *-(V)n* or by the residual locative *-(V)(t)t* as shown in (B3) above. Overabundance is possible exclusively with TOPOs which refer to settlements within the historical boundaries of the former Kingdom of Hungary.

---

76 For a definition of the notion of overabundance the reader is referred to Thornton (2012: 253–254).

Similarly, the use of the cases of the superior set with TOPOS is restricted to properly Hungarian geo-objects. In the case of foreign cities, the use of the interior set is compulsory. In this way it is possible to distinguish minimal pairs like inessive *Velencé-**ben*** 'in Venice (Italy)' from superessive *Velencé-**n*** 'in Velence (Hungary)' (Tompa 1972: 115–116). The distinction of (historically) national vs foreign TOPOS does not suffice to describe the range of variation because genuinely Hungarian TOPOS inflect differently from each other, too. There are phonological and morphological conditions which determine whether a given Hungarian TOPO takes the interior cases or those of the superior set. It is important to understand that nothing similar exists in the domain of COMMS or ANTHS. The choice of cases is completely independent of the phonological and morphological properties of the stem.[77] For TOPOS, however, the situation is very different.

According to Tompa's (1972: 116) summary treatment, morphologically complex TOPOS which have the structure of a compound whose rightmost member (= head) is a geo-classifier[78] like *-falu* 'village' always inflect in the interior cases because the head of the compound is identical with a COMM (in this case *falu* 'village') (but cf. below). For TOPOS which do not meet this morphological criterion, the phonological quality of the stem-final segment is decisive. Stems ending in /i/, /j/, /l/, /r/, /m/, /n/, /ɲ/ (i.e., stem-final segments with the properties [+high&+front] or [+sonorant]) usually combine with the interior cases. Violations of this rule are said to be absolutely exceptional (Bánhidi et al. 1975: 110). All other stems take the superior cases as shown in Figure 29.

What Figure 29 does not clarify is the allomorphy of the old locative *-t(t)* and the inessive. It is understood that the overabundance is attested only with TOPOS which refer to geo-objects within the historical boundaries of the erstwhile Kingdom of Hungary. The obsolete locative is admissible also on compound TOPOS such as *Székesfehérvár* → locative *Székesfehérvár-t* ~ *Székesfehérvár-ott* ~ superessive *Székesfehérvár-**on*** 'in Székesfehérvár' (Tompa 1972: 100). Interestingly, the compound structure of *Székesfehérvár* with the final head *vár* 'castle' does not impede the use of the superessive although the above rule requires the use of the inessive. Not only is *Székesfehérvár* a compound but it also has a stem-final rhotic sonorant /r/. Thus, there are two reasons for this TOPO to take the interior cases. As a matter

---

[77] With this statement we do not deny that there is morphophonological conditioning in the declension of Hungarian nouns. However, vowel-harmony and the stem-final segments determine only which of several allomorphs of one and the same case is employed. The variation observed with TOPOS is different from this scenario insofar as the properties of the stem are decisive as to which of several distinct case categories is admissible. This variation is certainly not the same as the above garden-variety of allomorphy.

[78] We employ this term according to the definition given by Anderson (2007: 106–107).

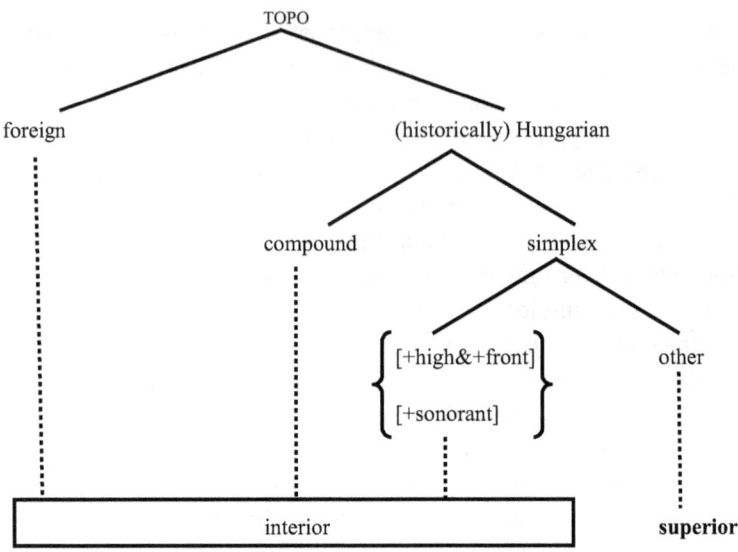

Figure 29: Hierarchy of conditions determining the use of spatial cases with TOPOs in Hungarian.

of fact, however, every TOPO with final *-vár* takes the superior cases provided reference is made to a settlement in the historically much larger territory of Hungary. The two examples in (65) show that this rule applies to *Kolozsvár* (< *Kolozs* 'Cluj' + *vár* 'castle') – the Hungarian name of modern *Cluj* which is situated on Romanian territory today, whereas *Újvár* (< *új* 'new' + *vár* 'castle') – today *Nové Zámky* (< *nové* 'new:PL' + *zámky* 'castle:PL' in Slovakia – is inflected as a foreign TOPO although both of the cities once belonged to the Kingdom of Hungary (Boisserie 2008: 117–119).

(65) Hungarian [Uralic, Hungaric]
    (a) superior cases (Mikesy 1978: 188)
        Parasztruhába   *öltözött*   *és*   *elindult*
         farmer's_dress:POR.3SG:INESS dress:PRET and depart:PRET
        [*Kolozsvár-ra*], [*Kolozsvár-on*] *a*   *bíró*   *háza*   *felé.*
        [Cluj-SBLTV]   **[Cluj-SUPESS]**   DET judge house:POR.3SG towards
        'He dressed as a farmer and left **[for Cluj]**, **[in Cluj]** (he went) to the judge's place.'
    (b) interior cases (Mikesy 1978: 283)
        Jól   *érezte*   *magát*   **[*Újvár-ban*],**   *egész*   *nap*   *csak*
         well   feel   self:ACC **[Nové_Zámky-INESS]** entire   day   only
        aludt
         sleep:PRET
        'Feeling well in **[Nové Zámky]**, he only slept for an entire day [...].'

On account of this piece of evidence, the rules put forward in Figure 29 must be considered to be default rules which can be overruled by more specific rules.

In addition to the possibility of distinguishing homophonous foreign TOPOs and national TOPOs from each other by way of employing different sets of spatial cases, Tompa (1972: 116) also mentions another common practice which exploits the availability of interior and superior sets of cases. Many of the major administrative units of the national territory of Hungary – the so-called *(vár)megyék* 'districts; counties' (singular: *megye*) – are named after their principal urban centre. This means that a TOPO like *Nógrád* as such is ambiguous because it could refer either to the city or to the district of the same name. However, the use of the spatial cases is regulated so that the cases of the superior set can refer only to the city whereas those of the interior set refer exclusively to the district. Thus, there are minimal pairs like superessive *Nógrád-on* 'in (the city of) Nógrád' vs inessive *Nógrád-ban* 'in (the district of) Nógrád' (Tompa 1972: 116). No explanations are given as to cases of conflict of this kind in which the name of the city also requires the use of the interior set of spatial cases.

More generally, the names of extended territories take the interior set of spatial cases. This applies to almost all names of countries like *Ausztria* 'Austria' → inessive *Ausztriá-ban* 'in Austria', illative *Ausztriá-ba* 'to Austria', elative *Ausztriá-ból* 'from Austria'. There is a notable exception to this rule, viz. the Hungarian designation of Hungary itself (= *Magyarország*) which usually requires the use of the spatial cases of the superior set (Bánhidi et al. 1975: 110). Thus, the sublative is used on the name of this country in example (66) whereas the names of other countries would appear in the illative in the same construction.

(66) Hungarian [Uralic, Hungaric] {Bud 26}
 Pontosan miért is nem tudsz visszamenni
 exactly why only NEG be_able:2SG back:go:INF
 [Magyarország-ra]?
 [Hungary-SBLTV]
 'Why exactly can't you just go back **[to Hungary]**?'

This exception is all the more interesting as *Magyarország* is a transparent binary compound whose constituents are *magyar* 'Hungarian' and *ország* 'country, state'. As our example (63b) above shows, there are names of several other countries which share this compositional structure with *Magyarország*. This structural parallel notwithstanding, all other names of countries of this kind can be inflected only in the cases of the interior set. In Table 16, we illustrate this contrast.

Table 16: The differential use of the interior and superior sets of spatial cases with names of countries in Hungarian.

country	Place	Goal	Source
Germany	inessive	illative	elative
Németország	Németország-**ban**	Németország-**ba**	Németország-**ból**
Magyarország	Magyarország-**on**	Magyarország-**ra**	Magyarország-**ról**
Hungary	superessive	sublative	delative

This morphological difference cannot be explained with reference to the formal properties of the stems to which the case-suffixes attach since the stems are identical. Therefore, there must be a different reason for the choice of the set of spatial cases. What strikes the eye is the correlation of the superior set and TOPOs which refer to properly Hungarian geo-objects. Wherever the superior set is given preference over the interior set we are dealing with names of places which are located on historical Hungarian territory or designate the country of Hungary itself. It is tempting to assume that the superior set of cases is associated with a certain degree of "empathy" in the sense that if the three spatial relations Place, Goal, and Source are expressed exclusively by cases of the superior set they epiphenomenally also imply that the geo-object is properly Hungarian. In connection to this "empathetically" motivated distribution of the sets of spatial cases, we might also invoke the diachronic age of the residual locative which is only licit on TOPOs which meet the criterion of being located within the boundaries of the erstwhile Kingdom of Hungary. Trivially, this locative is old, older than the other spatial cases. It is unsurprising therefore that its domain is limited to the names of cities whose foundation dates back to the earliest periods of the history of Hungary. In a way, the old age of these TOPOs presupposes a high degree of "empathy" or familiarity with the places thus designated on the part of the speakers.

As an aside we also mention streetnames. In Vilmos Kondor's novel *Budapest Novemberben*[79] streetnames abound, many of which are taken from the city-map of Vienna. A distinction is made in correlation to the size of the street. To express general location, the cases of the superior set are employed with major streets or squares (whose names involve the geo-classifiers -*straße* 'street' or -*platz* 'square' as, e.g., *a Klosterneuburger Strassé-n* 'on Klosterneuburger Straße' and *a Gauss-platz-on* 'on Gauss-Square' {Bud 47}) whereas minor urban geo-objects preferably – but not exclusively – take the interior cases (e.g. those which involve the geo-classifier -*gasse* 'alley' like *a Bankgasse-ba* '**into** Bankgasse' {Bud 70}, but cf. *a*

---

[79] We are grateful to Maike Vorholt and her Hungarian teacher Judith Hock for kindly letting us use a copy of this novel for the purposes of this investigation.

Leopoldgassé-n 'on Leopoldgasse' {Bud 21}). Whether or not there is a consistent pattern needs to be studied separately. For the purpose of this investigation, it suffices to hypothesise that Hungarian microtoponyms behave similarly to Hungarian macrotoponyms in the sense that they select one of two competing sets of spatial cases to express general location.

The Hungarian situation is not absolutely straightforward. There are uncertainties as well as potential contradictions in the above sketch. These problems notwithstanding, the conclusion can be drawn that Hungarian TOPOS do not behave the same morphologically as Hungarian COMMS and sundry classes of nouns. In Table 17 we provide a still rather coarse-grained schematic summary of the properties on the basis of which it is possible to differentiate between the three classes of nouny units in Hungarian. The plusses indicate that a given feature applies to the class at issue. Grey shading additionally highlights those cells which host a plus sign.

Table 17: Distinctive features of classes of nouns in Hungarian.

criteria	TOPOS	ANTHS	COMMS
number distinction	–	+	+
only one set for general location	+	–	–
pseudo-defectiveness	+	–	–
overabundance	+	–	–
residual locative	+	–	–

Table 17 is by no means suggestive of the tacitly expected continuum, the two extremes of which are occupied, on the one hand, by COMMS and, on the other hand, by TOPOS whereas ANTHS are situated in between. We would have assumed that TOPOS and ANTHS share the invariability as to the grammatical category of number. However, ANTHS even boast of two different kinds of plural. On the one hand, there is the regular plural in *-(V)k* which is used to refer to several individuals of the same name. Moreover, there is also the associative plural in *-ék* which is used to refer to a group of persons who belong to the same family. In this way the family name *Kovács* can have two distinct plurals, namely *Kovácsok* (= several not necessarily related persons who bear the same last name) and *Kovácsék* (= the family *Kovács*). The associative plural is also possible with COMMS if they refer to families as, e.g., *orvos* 'doctor' → *az orvosék* 'the doctor and his/her family' (Mikesy 1978: 86). The monopoly of one set of spatial cases in the domain of general location, overabundance, and the presence of the residual locative are traits which are characteristic exclusively of TOPOS. In Table 17, there are no properties which TOPOS have in common with COMMS. ANTHS, however, display similarities with COMMS on all of the parameters. Thus, it is obvious that TOPOS are special and can be considered, if not as a fully distinct word-class of its own, as a sub-class on the margins of the major word-class of nouns. ANTHs are much more like COMMs than they resemble TOPOs morphologically.

Figure 30 features the split rule we postulate for the choice of the sets of locative cases to be used to express general location.

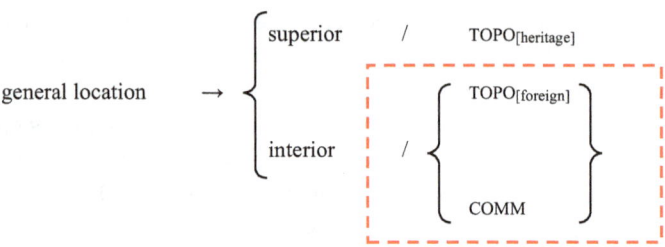

Figure 30: Split rule for general location in Hungarian.

The conditions are grossly simplified insofar as we assume the distinction between inherited and foreign TOPOs. The latter class contains, however, also those inherited Hungarian TOPOs which behave like the genuinely foreign TOPOs. Simplifications of this and similar kinds will be employed also for the remaining cases to be discussed in Section 5.

Special rules are required to describe the properties of TOPOs in the domain of the spatial categories. It goes without saying that the picture painted for Lezgian and that painted for Hungarian are not absolutely identical. However, it is worth taking note of some of the similarities. In both languages, the superior-interior split affects only local TOPOs. Foreign TOPOs are inflected homogeneously according to the majority pattern. Local TOPOs, on the other hand, are divided into two classes, namely those which combine with superior cases as opposed to those which require the interior cases. The distinction of two classes is restricted to TOPOs which refer to settlements (cities, villages) and probably also to streetnames whereas, except *Magyarország* 'Hungary', names of extended territories such as counties, countries, and continents are inflected homogeneously. It can be speculated whether or not there are different strategies of conceptualisation behind the above distinction. Where the interior cases are employed, the geo-object is conceived of as a container as opposed to the surface-like character of those geo-objects which call for the use of the superior set of spatial cases. Tentatively one may hypothesise that the differentiation is made on the grounds of familiarity, in a manner of speaking, whereas outside the home territory, a fine-grained differentiation is no longer feasible because of the lack of direct experience of the speakers with "the big world". On the other hand, streetnames do not fit this scenario particularly well. Moreover, it is also interesting to see that both languages exploit primarily the superior-interior distinction although their system of spatial categories comprises more sets of cases. The next two sections will check whether or not these parallels also count for other members of the Uralic language family.

5.2.2 Finnish

As in the Hungarian case, the exact size of the case-system of Finnish is a controversial issue. The estimates oscillate between a minimum of eleven and a maximum of fifteen cases. We need not dwell on this question because those cases which are of interest for our study are always accepted as integral parts of the inventory of Finnish cases.[80] The system of spatial cases of Finnish differs crucially from that of its sister-language Hungarian since in Finnish, there are only two sets, namely the interior and the exterior set. The superior set of Hungarian has no independent equivalent in the declension of Finnish nouns. However, the cases of the exterior set do not only express vicinity/approximation but also superiority (Karlsson 1984: 132–135). There are altogether six spatial cases in Finnish, namely
- in the interior set: inessive, illative, elative, and
- in the exterior set: adessive, allative, ablative.

Finnish COMMs regularly inflect for these cases as the COMM *talo* 'house' shows (cases of the singular only):
- *talo* 'house' → inessive *talo-ssa* '**in** the[81] house', illative *talo-on* '**into** the house', elative *talo-sta* '**out of** the house', adessive *talo-lla* '**at/on** the house', allative *talo-lle* '**(on)to** the house', ablative *talo-lta* '**(down) from** the house' (Karlsson 1984: 259).

Since several of the spatial cases also have more abstract, i.e., grammatical functions (such as the adessive as marker of the possessor in predicative possession[82]), it is unsurprising to see that ANTHs inflect like COMMs in Finnish. In (67), we illus-

---

80 For an analysis of the interaction of spatial cases and adpositions in Finnish, cf. Lestrade (2010a).
81 Finnish is a language without articles. The inflected word-forms of the nouns can have a definite or an indefinite reading. For consistency, we use the definite article in the translations (except otherwise explained).
82 As in {HP I Finnish, 7} *[Herra ja rouva Dursley-lla] oli pieni poika nimeltä Dudley. . . '[Mr and Mrs Dursley] had a little boy named Dudley [. . .].'* with adessive inflection on the family name Dursley in the Finnish version.

trate the inflectional potential of ANTHs in Finnish with an example each of the interior and the exterior set of cases.

(67) Finnish [Uralic, Finnic] – spatial cases with ANTHs
    (a) illative {HP I Finnish, 295}
        Harry kääntyi [Hermione-en] päin.
        Harry turn:PRET [Hermione-ILL] towards
        'Harry turned [**towards** Hermione].'
    (b) allative {HP I Finnish, 289}
        Meidän täytyy mennä kertomaan [Dumbledore-lle].
        1PL.GEN must:3SG go:INF1 report:INF3:ILL [Dumbledore-ALL]
        'We must go to report [**to** Dumbledore].'

TOPOs behave differently from both COMMs and ANTHs insofar as they split into two groups according to which of the sets of spatial cases a given TOPO combines with in order to express general location. There is the majority of the TOPOs which exclusively take the cases of the interior as opposed to a minority of TOPOs which can combine only with the cases of the exterior set (Karlsson 1984: 137–138). For each of the spatial relations Place, Goal, and Source, we provide a pair of sentential examples which illustrate the differential behaviour of Finnish TOPOs. In (68), the static relation of Place is documented whereas the relation of Goal is exemplified in (69) and that of Source in (70).

(68) Finnish [Uralic, Finnic] – Place
    (a) inessive (Mohtaschemi-Virkkunen 1970: 9)
        *Miksi ne eivät asu [**Helsingi-ssä**]?*
        why 2PL NEG:2PL live [**Helsinki-INESS**]
        'Why don't you live [**in Helsinki**]?
    (b) adessive (Karlsson 1984: 2003)
        *Veljeni on opiselemassa [**Tamperee-lla**].*
        brother:POR.1SG be.3SG study:INF3:INESS [**Tampere-ADESS**]
        'My brother studies [**in Tampere**].'

(69) Finnish [Uralic, Finnic] – Goal
    (a) illative (Mohtaschemi-Virkkunen 1970: 16)
        *En, minä matkustan [**Vaasa-an**].*
        NEG:1SG 1SG travel:1SG [**Vasa-ILL**]
        'No, I travel [**to Vasa**].'

(b) allative (Karlsson 1984: 121)
   Rakennamme   tehtaan   *[Tamperee-lle]*.
   build:1PL   factory:GEN   **[Tampere-ALL]**
   'We build a factory **[in Tampere]**.'[83]

(70) Finnish [Uralic, Finnic] – Source
   (a) ablative (Mohtaschemi-Virkkunen 1970: 16)
       Kiitos,   mutta   minä   en   tule   *[Tamperee-lta]*.
       thanks   but   1SG   NEG:1SG   come   **[Tampere-ABL]**
       'Thank you, but I do not come **[from Tampere]**.'
   (b) elative (Mohtaschemi-Virkkunen 1970: 17)
       En   minä   tule   *[Turu-sta]*.
       NEG:1SG   1SG   come **[Turku-ELA]**
       'I don't come **[from Turku]**.'

These examples are indicative of a bipartition of TOPOS. In the realm of spatial categories, TOPOS like *Helsinki, Turku,* and *Vaasa* inflect differently from TOPOS like *Tampere, Rauma,* and *Rovaniemi*. When it comes to encoding general location, the latter class of TOPOS lacks the interior set of spatial cases whereas the more numerous former class hosts those TOPOS which do not inflect in the exterior cases (cf. Table 18).

Table 18: Spatial cases of the interior and exterior set of Finnish TOPOS and COMMS – expression of general location.

set	case	TOPOS		COMM
		Helsinki	*Rauma*	*kauppa* 'shop'
interior	inessive	Helsingi-ssä	---	kaupa-ssa
	illative	Helsingi-in	---	kauppa-an
	elative	Helsingi-stä	---	kaupa-sta
exterior	adessive	---	Rauma-lla	kaupa-lla
	allative	---	Rauma-lle	kaupa-lle
	ablative	---	Rauma-lta	kaupa-lta

---

83 Literally: 'We build a factory **[unto Tampere]**.' *Rakentaa* 'to build' belongs to the class of verbs which are indicative of a change of state. These verbs require the spatial adverbial adjunct to be marked for a dynamic spatial relation – in this case, depending on the TOPO, either the illative or the allative (Karlsson 1984: 137).

What Table 18 shows additionally is that their internal differences notwithstanding, the TOPOs have an important property in common which sets them apart from COMMs (and ANTHs). This distinctive property is the impossibility to morphologically distinguish different dimensions of general location. Where COMMs (and ANTHs) are equipped with six different spatial cases, TOPOs only allow for three spatial cases.

Finnish has a sizable class of postpositions, many of which can be employed to describe spatial situations in more detail than is possible with the dedicated case-markers of general location. Superior location without contact, for instance, is expressed periphrastically by a postpositional phrase whose head is the postposition *y-* 'over, above' which is inflected for case (and governs the genitive on the complement noun) (Stoebke 1968: 108–110). In example (71), the postposition comes in the shape of the adessive *yllä* 'over, above' to express a static relation of Place (cf. (66a) above).

(71) Finnish [Uralic, Finnic] {HP I Finnish, 22} – Place [+superior&-contact][84]
 Poitsu nukahti jossain **[Bristol-in yllä].**
 boy take_a_nap:PRET somewhere **[Bristol-GEN over]**
 'The boy fell asleep somewhere **[over Bristol]**.'

This strategy is also common with COMMs and ANTHs. Note that *yllä* 'over, above' and related postpositions do not compensate for the entire range of functions of the cases of the exterior set. Thus, the gaps in the paradigms shown in Table 18 can be filled only in part by periphrastic constructions. However, as in the Hungarian case, defectiveness is not a general trait of the declension of TOPOs in Finnish.

In point of fact, most of the putative gaps in the paradigm of TOPOs can be filled because there are constructions which require the use of a given spatial case other than those employed for the purpose of general location. Moreover, there are postpositions and verbs which govern spatial cases which are excluded from the realm of general location. By way of example, we present three constructions in which the TOPOs *Helsinki* and *Tampere* replace each other in a given slot. As we know from examples (68)–(70), for the purpose of expressing general location, *Helsinki* opts for the spatial cases of the interior set as opposed to *Tampere* which combines with those of the exterior set. In (72)–(74), these rules are not obeyed.[85]

---

[84] This is an utterance of the character Hagrid whose speech is clearly marked as nonstandard lexically and phonologically. The nonstandard status of the utterance, however, does not affect the postpositional phrase.
[85] The sentences (72)–(74) are modfications of sentential examples provided in Karlsson (1984: 124–138 and 232–234). Their acceptability has been checked with a Finnish native-speaker linguist, Eeva Sippola, whose kind support is herewith gratefully acknowledged.

Example (72), for instance, shows that a verb like *kertoa* 'to report' takes a complement NP in the elative irrespective of the behaviour of the same NP in the domain of general location. This means that the TOPO *Tampere* hosts a case which is banned from those constructions in which *Tampere* is involved in spatial relations of the general-location kind.

(72) Finnish [Uralic, Finnic] – verb government

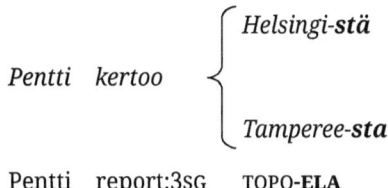

Pentti   report:3SG   TOPO-ELA

'Pentti reports **on** Helsinki/Tampere.'

In (73), on the other hand, it is the TOPO *Helsinki* that has to take the adessive inflection because this is what the construction of predicative possession requires of the NP which fulfils the function of possessor.

(73) Finnish [Uralic, Finnic] – predicative possession

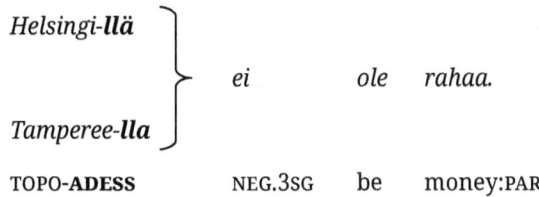

TOPO-**ADESS**   NEG.3SG   be   money:PART

'Helsinki/Tampere has no money.'

Interestingly, our Finnish informant (Eeva Sippola, personal communication) found fault with the illative inflection on the TOPO *Tampere* in Example (74) where the postposition (*päin* 'towards') is supposed to govern the illative case on its complement (Karlsson 1984: 235). However, the allative was rated much more acceptable with the TOPO *Tampere*.

(74) Finnish [Uralic, Finnic] – postpositional phrase

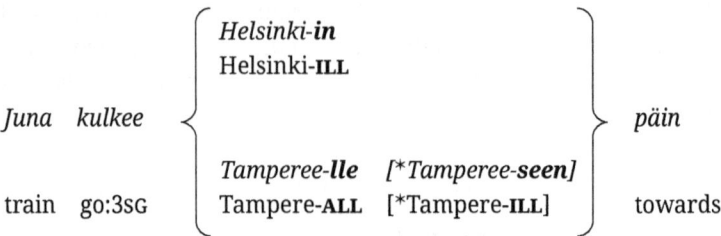

Juna kulkee ... *päin*
train go:3SG ... towards

{ Helsinki-**in** / Helsinki-ILL }

{ Tamperee-**lle** [*Tamperee-**seen**] / Tampere-ALL [*Tampere-ILL] }

'The train is going **to** Helsinki/**to** Tampere.'

The blocking of the illative government of *päin* 'towards' in combination with *Tampere* renders the picture of the morphological behaviour of Finnish TOPOS definitely more colourful because, in contrast to what we have seen in the case of Hungarian, one cannot sweepingly assume that an external head is always powerful enough to dictate the employment of a case that is not included in the set of spatial cases which is used for the purpose of general location. What is worth noting is that resistance against the case-assignment arises only in (74), i.e., in a context that involves a situation which is properly spatial. In contrast, Examples (72)–(73) reflect situations which are conceptually remote from purely spatial configurations synchronically (no matter how strong the spatial component was at the beginning of the grammaticalisation processes in the distant past). As shown in Example (75), the illative is even compulsory on *Tampere* if the TOPO is the internal argument of a verb like *rakastua* 'to fall in love'.

(75) Finnish [Uralic, Finnic] – illative government[86]
Viikonloppu-na   voi   rakastua   [Tamperee-seen].
weekend-ESS   be_abble:3SG   fall_in_love:INF   [Tampere-ILL]
'On weekends you can fall in love [**with Tampere**].'

The situation described in (75) is clearly not properly spatial. Therefore, the use of the illative is admissible with *Tampere*. There is thus a relatively strict division into genuinely spatial contexts and those which fail to meet the criteria of describing a spatial configuration.

The Finnish givens do not allow us to classify them as evidence of morphological defectiveness. Examples (72)–(73) and (75) are indicative of the ability of all TOPOS to also inflect for those spatial cases which are not used by them in the realm

---

[86] http://www.city.fi/opas/viikonloppuna+voi+rakastua+tampereeseen/8799

of general location. In contrast to Hungarian, there is not even pseudo-defectiveness. The absence of this phenomenon is clearly reflected by the distribution of **yes** and **no** in Table 19.

Table 19: Two different TOPO types / Finnish.

TOPO type	set	function	
		general location	other
A	exterior	yes	yes
	interior	no	yes
B	exterior	no	yes
	interior	yes	yes

In contrast to the distribution of sets of cases in Hungarian, there do not seem to be genuinely phonological conditions which determine the choice of spatial cases on Finnish TOPOs. What Hungarian and Finnish have in common nevertheless is the formal differentiation of foreign TOPOs and TOPOs "at home". As in Hungarian, the Finnish designations of geo-objects which are located abroad tend to accept only the cases of the interior set. This holds for the names of cities. In the case of names of countries, there is the notable exception of *Venäjä* 'Russia' which combines with the cases of the exterior set (Karlsson 1984: 138). In (76) we provide sentential examples of the use of the interior set of cases with names of countries.

(76) Finnish [Uralic, Finnic] – names of countries
    (a) inessive (Karlsson 1984: 142)
        Kesällä   *olen*   *[Suome-ssä]*.
        summer:ADESS be:1SG **[Finland-INESS]**
        'In the summer I will be **[in Finland]**.'
    (b) illative (Karlsson 1984: 142)
        Kesäksi   *lähden*   *[Suome-en]*.
        summer:TRANS travel:1SG **[Finland-ILL]**
        'Next summer I will go **[to Finland]**.'
    (c) elative (Karlsson 1984: 128)
        Kekkonen   *on*   *palannut*   *[Brasilia-sta]*.
        Kekkonen be.3SG return:PTCPL **[Brazil-ELA]**
        'Kekkonen has returned **[from Brazil]**.'

ANTHs behave very much the same as COMMs. The similarity of the two classes of nouns holds also in the domain of number marking. Where TOPOs normally cannot undergo pluralisation, ANTHs are pluralisable just as COMMs. Example (77) contains a family name that is in the plural to refer to all members of the family of that name.

(77) Finnish [Uralic, Finnic] {HP I Finnish, 7}
Heistä tuntui etteivät he kestäisi,
3PL:ELA feel:PRET that:NEG:PRET:3PL 3PL endure:COND
jos joku saisi tietää *[Pottere-i-sta]*.
if someone take:COND know:INF **[Potter-PL-ELA]**
'They felt that they could not endure if someone came to know **[about the Potters]**.'

In contrast to Hungarian, however, standard Finnish does not have a distinct associative plural. The absence of number distinctions is one aspect that separates TOPOS from other classes of nouns. Since, in the paradigm of Finnish TOPOS, there are no residual cases like the obsolete locative of Hungarian and since there is also no evidence of overabundance, the main difference of TOPOS and other classes of nouns is the specialisation of one set of spatial cases for general location in the paradigm of the former. This is shown in Table 20.

Table 20: Distinctive features of classes of nouns in Finnish.

criteria	TOPOS	ANTHS	COMMS
number distinction	–	+	+
one set for general location	+	–	–

As we have observed in connection to the Hungarian situation, the structural differences of TOPOS and other classes of nouns do not yield a continuum since there is no transition zone.

To sum up, we recapitulate a selection of our above observations in Figure 31 which is reminiscent of Figure 30 reflecting the Hungarian situation.

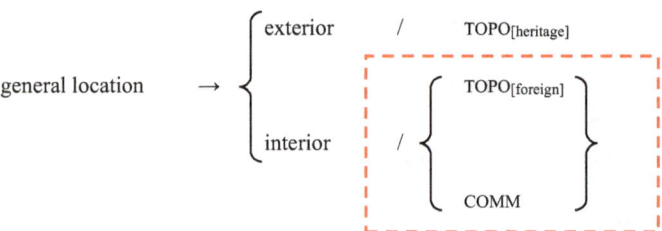

Figure 31: Split rule for general location in Finnish.

As in the Hungarian case, the indexes [heritage] and [foreign] simplify the intricate facts but the general picture is captured adequately by Figure 31. The exterior set of cases is restricted to a selection of names of settled places in Finland. The split of TOPOs as to which set of spatial cases is employed exclusively affects properly Finnish geo-objects in the case of cities, towns, and villages. *Venäjä* 'Russia' is the sole exception in the domain of names of countries and thus hardly impairs the impression of wholesale homogeneity of this class of TOPOs. The hypothesis according to which "empathy" is a factor to be taken account of if one wants to explain the distribution of different patterns of morphological behaviour of TOPOs receives some support also from Finnish since only places which are located in Finland are eligible for combining with the exterior cases. Discounting the differences on the micro-level, Hungarian and Finnish behave remarkably similarly. One might object that this similarity is unsurprising because we are dealing with genetically related languages. If the genetic argument is a valid argument, the situation in Estonian should be not much different from that of its two sister languages.

5.2.3 Estonian

In point of fact, the Estonian situation resembles that of the two major Uralic languages. First of all, there is also a sizable system of inflectional cases. According to Viitso (2003: 32) this system comprises fourteen distinct categories in both numbers. In addition to the terminative *-ni* (as in, e.g., *kivi* 'stone' → terminative *kivini* 'up to the stone'), we find two ternary sets of spatial cases as in Finnish, namely
- the interior set consisting of inessive, illative, and elative,
- the exterior set consisting of adessive, allative, and ablative.

Estonian COMMs are regularly inflected for all six of these cases as shown by the following word-forms from the paradigm of *hammas* 'tooth' (Viitso 2003: 39) – only word-forms of the singular are taken account of:
- *hammas* 'tooth' → inessive *hamba-s* '**in** the[87] tooth; illative-2 *hamba-sse* '**into** the tooth'; elative *hamba-st* '**out** of the tooth'; adessive *hamba-l* '**on** the tooth'; allative *hamba-le* '**onto** the tooth'; ablative *hamba-lt* '**from** the tooth'.

---

[87] As in the previous case of Finnish, we use the definite article in the English translation of the isolated inflected Estonian word-forms irrespective of the possibility of an indefinite reading of the same example.

The illative comes in two shapes, viz.
- illative-1 = -*sse* and
- illative-2 = -*de* ~ -*ha* ~ -*he* ~ -*hu* ~ Ø

which are often referred to as long illative and short illative. Viitso (2003: 40–41) assumes that phonologically conditioned allomorphy applies according to which the illative-1 requires a stem with a final vowel as in
- *maa* 'land, country' → illative-1 *maa-sse* '**into** the land'

whereas the various allomorphs of the illative-2 combine a variety of phonological conditions which we summarise here by way of glossing over a number of details. In the case of -*ha* ~ -*he* ~ -*hu*, the stem must be monosyllabic and host a final vowel. The allomorph that represents the illative-2 is selected in accordance with vowel harmonic preferences as in
- *öö* 'night' → illative-2 *ö-he* '**into** the night';
- *suu* 'mouth' → illative-2 *su-hu* '**into** the mouth'.

The allomorph -*de* occurs with stems which end "in a long vowel, diphthong or the consonant *l, n, r*" (Viitso 2003: 40) as in
- *joon* 'line' → illative-2 *joon-de* '**into** the line'.

The zero-allomorph of the illative-2 is employed mostly on polysyllabic stems such as
- *taevas* 'sky' → illative *taeva* '**into** the sky'.

Many COMMs allow for the coexistence of forms of illative-1 and illative-2 in their paradigm as is the case with
- *keel* 'tongue, language' → illative-1 *keele-sse* 'in the language' (based on the so-called vocalic weak-grade stem) ~ illative-2 *keel-de* 'in the languages'.

Beyond these morphophonological facts which are interesting by themselves already, what is even more worthy of note is Viitso's (2003: 41) contention according to which

> [t]he use of forms of illative-1 and illative-2 is not entirely free. When a noun has in its paradigm both illative forms, the short form has a tendency to be substituted for the interrogative adverb '*kuhu* 'where to' whereas the corresponding long form can be substituted for the pronoun '*millesse* 'in(to) what'.

This subliminal distinction is assumed to depend on verb government so that one may say
- *lähen kooli* 'I go **to** school' – with *kool* 'school' → illative-2 *kooli* 'into the school',

i.e., a zero-marked illative-2 based on the inflectional stem *kooli* – whereas **lähen kooli-sse* 'I go **to** school' is considered ungrammatical by Viitso (2003: 41) who also assumes that in the modern standard language, the allomorph *-de* of the illative-2 is expanding its domain to the detriment of the forms of the illative-1. For COMMs whose paradigms contain two formally distinct illatives, it can be assumed that they give testimony perhaps not only of overabundance but also of **overdifferentiation** if it is indeed possible to use the different illatives to answer questions which are based on different spatial interrogatives.

This summary of the system of allomorphy of the Estonian illative is necessary if one wants to determine to what extent TOPOs differ from COMMs morphologically in this Uralic language. Intuitively it makes sense to assume that TOPOs which inflect in the interior cases prefer the illative-2 over the illative-1 because one expects them to answer a virtual question which involves the interrogative adverb *kuhu* 'where to' in lieu of *millesse* 'into what'. This hypothesis will be checked further below.

From the extant descriptions it is evident that in Estonian (like in Finnish and Hungarian), TOPOs do not inflect for the entire set of spatial cases for the purpose of expressing general location. The names of settlements differ from those of countries, islands, and other geo-objects which as classes display a clear preference for the spatial cases of only one of the two sets as, e.g.,
- *Rootsi* 'Sweden' → inessive *Rootsi-s* '**in** Sweden', illative *Rootsi-sse* '**to** Sweden', elative *Rootsi-st* '**from** Sweden'.

Since variation is encountered only with names of settlements, we focus on this class of TOPOs. All foreign TOPOs behave according to one and the same pattern in Estonian insofar as they employ exclusively the spatial cases of the interior set. As to TOPOs which refer to settlements within the boundaries of Estonia, however, Tauli (1983: 99) claims that the majority of the Estonian TOPOs takes the cases of the interior set whereas a not negligible minority gives this privilege to the exterior set. In addition, there is also variation with individual TOPOs. This estimate is largely corroborated by the statistics in Erelt (2002).[88] On the basis of 1,236 TOPOs of Estonia, the following distribution emerges:

---

[88] Note that the 2013 online version of Erelt (2002) deviates occasionally (at least nine times) from the information which is given in the original print version, i.e., the two versions disagree as to which of the two sets of spatial cases is to be used with a given TOPO.

- with 763 TOPOS a majority of 62% opt for the interior set,
- a minority of 29% (= 357 TOPOS) inflects exclusively in the cases of the exterior set, and
- some 9% (= 116 TOPOS) are examples of free variation of both sets.

The distribution of the sets of spatial cases over inner-Estonian TOPOS is reflected schematically in Figure 32.

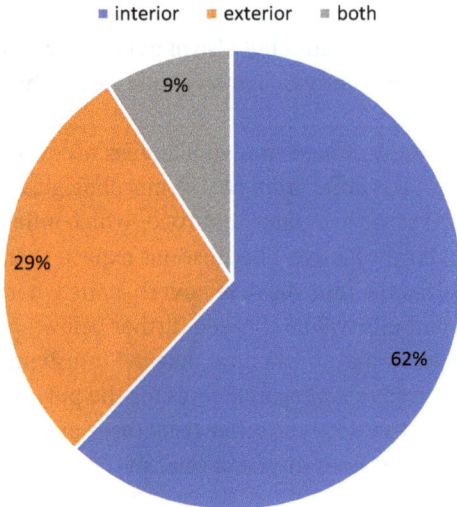

Figure 32: Competition of the sets of spatial cases with Estonian TOPOS.

Hennoste and Pajusalu (2009: 268) provide a statistic which also counts a majority of 61% of TOPOS which take the cases of the interior set. However, they report that 48% of the Estonian TOPOS are inflected for the cases of the exterior set. The total of 109% is caused by the authors' decision not to introduce a separate category of TOPOS which allow for the synonymous use of spatial cases of both sets in free variation. The relatively big share of TOPOS which inflect for exterior spatial cases can be explained as the logical consequence of the inclusion of TOPOS other than the names of settlements in Hennoste and Pajusalu's (2009) study.

In the descriptive-linguistic accounts, it is largely uncontroversial that for TOPOS which have the internal structure of compounds, the rightmost constituent (= head) determines the choice of the set of spatial cases in which the TOPO can be inflected when general location is to be expressed. What is striking about this fact is that these constituents which are identical with COMMS can be declined in both sets of spatial cases when used as syntactic words of their own. This means that the

restriction on the declinability of the compound-like TOPOs is not directly inherited from the morphological potential of their heads. A selection of heads of TOPO compounds and their preferences as to the set of spatial cases is given below. As heads of TOPO compounds these nouns are usually in genitive singular (Tauli 1983: 99):
- interior set: *-küla* 'village', *-linn(a)* 'town', *-metsa* 'wood', *-mõisa* 'estate', *-ranna* 'beach', *-saare* 'island', *-salu* 'forest', etc.
- exterior set: *-jõe* 'river', *-kivi* 'stone, rock', *-maa* 'land', *-pää* 'head', *-selja* 'ridge', etc.
- variable: *-jala* 'foot', *-järve* 'lake', *-nurme* 'field', *-soo* 'swamp', etc.

According to a deliberately naïve reading of this list, Estonian TOPOs with final *-kivi* should have no interior cases since they take those of the exterior set (namely adessive *-kivi-l*, allative *-kivi-le*, and ablative *-kivi-lt*). The COMM *kivi* 'stone, rock', however, also inflects for the cases of the interior set as is shown in (78).

(78) Estonian [Uralic, Finnic] {HP I Estonian, 253}
Olgu   peale,   räägime   [kivi-st].
be:3IMPTV at   speak:1PL [stone-ELA]
'Well then, let's talk [about the stone].'

In this example, the elative is required by the verb *rääkima* 'to speak, to talk (about)'. Since it is trivially also possible to talk about settlements, TOPOs occur as complements of the same verb. Those TOPOs which inflect for the interior set anyway pose no problem. However, how do those TOPOs behave which take the exterior set in the context of general location? As in Finnish and Hungarian, the external trigger is stronger than the morphological preferences the TOPOs display in the realm of general location.

With COMMs, the choice of inflectional case is not lexically determined but depends on the semantic distinctions one wants to make explicit. There is thus no free variation either. It makes a difference from which of the sets of spatial cases a word-form is picked, cf. (79)–(80).

(79) Estonian [Uralic, Finnic] – Place / COMM
    (a) inessive / COMM {HP I Estonian, 94}
        Millises   [maja-s]   su   vennad   on?
        which:INESS [house-INESS] your brother:PL be.3SG
        '[In] which [house] are your brothers?'

(b) adessive / COMM {HP I Estonian, 100}
Igal   **[maja-l]**   on   oma   üllas   ajalugu
each:ADESS **[house-ADESS]** be.3SG   POR.RFL   noble   history
'Each **[house]** has its own noble history...'

(80) Estonian [Uralic, Finnic] – Goal / COMM {HP I Estonian, 101}
    (a) illative / COMM
    Kuidas   täpselt   nad   meid   **[maja-de-sse]**   söölavad?
    how   exactly   3PL   1PL:PART **[house-PL-ILL]** sort:3PL
    'How exactly do they assign us **[to the houses]**?'
    (b) allative / COMM
    Ma   loodan,   et   igaüks   teist   teeb   au
    1SG   assume:1SG   SUBORD   each   2PL:ELA   earn:3SG   honor
    [maja-le],   mis   tema   omaks   saab.
    [house-ALL]   which   3SG   POR.self:TRANS   become:3SG
    'I assume that each of you earns honor **[for the house]** to which they belong.'

In (79), the COMM *maja* 'house' is used twice, viz. once each in the inessive and the adessive. Similarly, the same COMM occurs twice in (80) – once each in the illative and the allative. Arguably, the situations are not properly spatial. The pair of sentences in (79) is representative of possessive relations. This is especially obvious in the case of (79b) where the regular construction of predicative possession is employed which requires the possessor to be in the adessive. Example (79a) is less straightforward (and thus, perhaps, more spatial than (79b)) because a context-free reading could be that of an inquiry as to the whereabouts of the brothers. However, the question posed by Harry Potter to his new acquaintance Ron Weasley can be rephrased alternatively as *To which house do your brothers belong?*, i.e., the sense is that of a so-called belong-construction. In (80a), the distribution of pupils over the four houses of the school can be considered an event in which the houses function as recipients whereas in (80b) the houses benefit from the behaviour of their members, i.e., a benefactive relation applies. There is no reason to assume that TOPOS are excluded apriori from constructions as those illustrated in (78)–(80) above. By way of proof, we provide examples of the inflection of Estonian TOPOS in spatial cases which are blocked for them in properly spatial contexts. In (81), the TOPOS are representative of the class whose members take the exterior set to express general location whereas (82) features TOPOS which use the cases of the interior set for the same function.

(81) Estonian [Uralic, Finnic] – interior cases
    (a) elative governed by noun[89]
        Hommikul oli meeste hommikusöök kus
        morning:ADESS be:PRET.3SG man:GEN.PL breakfast where
        oli presentatsioon [Ahja-st] ja Eesti Metodisti
        be:PRET.3SG presentation **[Ahja-ELA]** and Estonian Methodist
        Kirikust
        Church:ELA
        'In the morning there was the men's breakfast during which there was a presentation **[on Ahja]** and the Estonian Methodist Church.'
    (b) illative goverened by verb[90]
        mis [Jõgeva-sse] puutub siis kardetavasti me räägime
        what **[Jõgeva-ILL]** touch:3SG then regrettably 1PL speak:1PL
        samast kohast
        same:ELA place:ELA
        'As to **Jõgeva**, we regrettably seem to speak about the same place.'

(82) Estonian [Uralic, Finnic] – exterior cases
    (a) allative goverened by verb[91]
        Valitsus annab [Orissaare-le] 6,3 miljonit
        government give:3SG **[Orissaare-ALL]** 6.3 million:PART
        'The government gives 6.3 million **[to Orissaare]**.'
    (b) adessive (possessor in predicative possession construction)[92]
        Kahjuks puudub [Orissaare-l] teeäärne pilkupüüdev
        unfortunately lack:3SG **[Orissaare-ADESS]** kerbside eye_catching
        reklaam
        advertisement
        'Unfortunately **[Orissaare]** lacks an advertisment at the kerbside to attract attention.'

As in the previous Finnish and Hungarian cases, the restriction of TOPOs to one of the sets of spatial cases is limited to the expression of general location. In all other contexts, the TOPOs behave similarly to COMMs and ANTHs whose capacity to inflect for all kinds of cases can be gathered from the examples in (83).

---

89 https://jarereisib.wordpress.com/2008/01/22/poodides-ja-kodudes/
90 http://foorum.bmwclassic.ee/viewtopic.php?f=1&t=56&start=20
91 http://w3.ee/openarticle.php?id=350484&lang=est
92 https://www.meiemaa.ee/index.php?content=artiklid&sub=5&artid=49687&selyear=2015&selmonth=2

(83) Estonian [Uralic, Finnic] – ANTHS
(a) elative – {HP I Estonian, 22}
Dursleyd rääkisid **[Harry-st]** sageli nii...
Dursley:PL speak:COND:3PL **[Harry-ELA]** frequently thus
'The Dursleys used to speak **[about Harry]** often like this...'
(b) ablative – {HP I Estonian, 29}
Olles küsinud **[Harry-lt]** vihaselt, kas ta
be:PTCPL ask:PTCPL **[Harry-ABL]** anger:ELA, INTERR 3SG
tunneb seda meest...
know:3SG DEM.PROX:PART man:PART
'After having asked angrily **[of Harry]**, whether he knew this man...'

In (84) we contrast the use of the elative and the ablative to express the spatial relation of Source with different TOPOS. The different morphological cases are synonymous since they share the same function, namely that of indicating that a given human settlement is the point of departure of a motion event.

(84) Estonian [Uralic, Finnic] – Source / TOPOS
(a) elative / foreign TOPO (Lutkat-Lõik and Hasselblatt 2005: 67)
Laev tuleb **[Helsingi-st]**.
ship come:3SG **[Helsinki-ELA]**
'The ship comes **[from Helsinki]**.'
(b) elative / place in Estonia (Lutkat-Lõik and Hasselblatt 2005: 69)
Kas nad tulevad **[Tartu-st]**?
INTERR 3PL come:3PL **[Tartu-ELA]**
'Do they come **[from Tartu]**?'
(c) ablative / place in Estonia (Lutkat-Lõik and Hasselblatt 2005: 110)
Tulen teisipäeval **[Võsu-lt]** tagasi.
come:1SG Tuesday:ADESS **[Võsu-ABL]** back
'I'll come back **[from Võsu]** on Tuesday.'

In connection with TOPOS, speakers of Estonian must know which of the two morphological cases is admissible with which TOPO. This necessity does not generally arise with COMMS.

In Table 21 we survey the situation as it is reported for TOPOS which are monomorphic. The names of the cities of London, Tartu, and Kudina are compared to the COMM *ase* 'place' as to their susceptibility to the case distinctions of the interior and the exterior set. The word-forms of the COMM are based on the genitive singular *aseme* 'of the place' which is the regular inflectional stem for many of the case-forms of Estonian nouns.

5.2 Uralic parallels

Table 21: Spatial cases of the interior and exterior set of Estonian TOPOS and COMMS – expression of general location.

set	case	TOPOS			COMM
		foreign	in Estonia		
		London	Tartu	Kudina	*ase* 'place'
interior	inessive	*Londoni-s*	*Tartu-s*	---	*aseme-s*
	illative	*Londoni-sse*	*Tartu-sse*	---	*aseme-sse*
	elative	*Londoni-st*	*Tartu-st*	---	*aseme-st*
exterior	adessive	---	---	*Kudina-l*	*aseme-l*
	allative	---	---	*Kudina-le*	*aseme-le*
	ablative	---	---	*Kudina-lt*	*aseme-lt*

The picture that emerges from Table 21 is already familiar. Where the COMM yields six distinct word-forms, the three TOPOS have to make do with three distinct word-forms each. On top of this, *London*, *Tartu*, and *Kudina* behave differently from each other insofar as the former two inflect exclusively for the cases of the interior set whereas *Kudina* can be declined only in the cases of the exterior set. However, as shown above, we cannot consider them candidates for the status of defective paradigms. Table 22 shows that ANTHS and COMMS share the ability to inflect in two sets of spatial cases for the purpose of general location – a property which is alien to TOPOS.

Table 22: Distinctive features of classes of nouns in Estonian.

criteria	TOPOS	ANTHS	COMMS
two spatial sets	–	+	+
one spatial set	+	–	–

The system of spatial cases as it manifests itself in the domain of Estonian TOPOS closely resembles that of Finnish. This unsurprising fact is revealed in Table 23 which is almost an exact replica of Table 19 above.

The only but nevertheless striking difference is the existence of Class C in Estonian. The members of this class differ from those of Classes A–B insofar as they give evidence of the use of both sets of spatial cases under general location. However, in contrast to COMMS and ANTHS, this possibility is supposed to be a case of synonymy or free variation of the two sets, meaning: where the interpretation of say, the inessive and the adessive of a COMM invokes two different spatial situations, the two cases can be used interchangeably on TOPOS of Class C without altering the spatial situation in any way.

Table 23: Two different TOPO types / Estonian.

TOPO type	set	function	
		general location	other
A	exterior	yes	yes
	interior	no	yes
B	exterior	no	yes
	interior	yes	yes
C	exterior	yes	yes
	interior		yes

As mentioned above, the illative of COMMs is subject to allomorphy so that many nouns give evidence of overabundance which involves a long form and a short form of the illative. TOPOs do not behave much differently from COMMs. It is difficult to formulate any general rule which would cover the majority of the TOPOs, some of which allow only one of two possible forms of the illative whereas others tolerate the coexistence of the two allomorphs according to the model of the COMMs. The name of the village *Salutaguse*, for instance, provides evidence of two illatives, namely illative-1 *Salutaguse-sse* 'into Salutaguse' and illative-2 *Saluta-ha* 'into Salutaguse', as opposed to *Mäetaguse* which only allows the illative-1 *Mäetaguse-sse* 'into Mäetaguse' although both of the TOPOs involve the head *-taguse* (< *tagune*) 'back (part of)'. In (85)–(87), we provide sentential examples of the variation of interior and exterior spatial cases with the TOPO *Mäetaguse*. Note especially the expression of the static spatial relation of Place in (85b) and the dynamic spatial relation of Source in (87b). In Example (85b) there is an enumeration of five TOPOs, all of which encode the spatial relation of Place – however, by means of different morphological cases. Similarly, in (87b) Source is encoded on three TOPOs, namely twice by the ablative case and once by the elative case.

(85)   Estonian [Uralic, Finnic] – Place
   (a)   inessive[93]
       [Mäetaguse-s]   *tervitas*   *päris*   *tuuline*   *ilm*
       [Mäetaguse-INESS]   greet:PRET.3SG   really   windy   weather
       '**[In Mäetaguse]** really windy weather greeted...'

---

[93] http://www.ccrotamobilis.ee/emv-eraldistardis-esimene-kuldmedal-klubi-ajaloos/

(b) adessive[94]

    Gennadi   on   koolipinki   nühkinud   nii
    Gennadi be.3SG school_bench rub:PTCPL not_only
    ***[Mäetaguse-l]**,   [Vändra-s],   [Tõrva-s],   [Kose-l]*
    [Mäetaguse-ADESS] [Vändra-INESS] [Tõrva-INESS] [Kose-ADESS]
    kui ka   [Kehra-s].
    as   also [Kehra-INESS]
    'Gennadi has not only rubbed the school-bench **[in Mäetaguse]**, [in Vändra], [in Tõrva], [in Kose] but also [in Kehra].'

(86) Estonian [Uralic, Finnic] – Goal
    (a) illative[95]

        ta   olla   kavatsenud ka   üht   Hernhuti-meest
        3SG   be:INF plan:PTCPL   also   one:PART Herrnhut-man:PART
        ***[Mäetaguse-sse]** kutsuda*
        [Mäetaguse-ILL] call:INF
        'Supposedly he had planned to invite also a man from Herrnhut **[to Mäetaguse]**...'

    (b) allative[96]

        *aga   **[Mäetaguse-le]**   me   jõudsime*
        but   **[Mäetaguse-ALL]** 1PL   reach:PRET.1PL
        'But we came **[to Mäetaguse]**...'

(87) Estonian [Uralic, Finnic] – Source
    (a) elative[97]

        1857.   aastal   kolis   koos   perekonnaga
        1857   year:ADESS move_house:PRET.3SG together family:COM
        ***[Mäetaguse-st]**   Kalina   mõisasse.*
        [Mäetaguse-ELA] Kalina.GEN estate:ILL
        'In the year 1857, he moved together with his family **[from Mäetaguse]** to the Kalina estate.'

---

94 http://kes-kus.ee/akadeemiline-keskus-keskusi-lugudesari-eesti-akadeemikuist-tana-vaatame-akadeemik-gen-nadi-vainikko-tegevuse-kaudu-mida-tahendab-matemaatikas-termin-sudamlik/
95 http://www.folklore.ee/radar/story.php?area=J%F5hvi&id=294
96 http://www.kuula.ee/blog/
97 http://www.geni.com/people/Jaan-T%C3%B5nurist/6000000015270991848

(b) ablative[98]

Õpilasi	tuleb	meile	ka	Jõhvi	ümbrusest
pupil:PART.PL	come:3SG	1PL:ALL	too	Jõhvi.GEN	surroundings:ELA
ja	**[Mäetaguse-lt]**,	[Ahtme-st]	ja	isegi	[Sillamäe-lt]
and	**[Mäetaguse-ABL]**	[Ahtme-ELA]	and	even	[Sillamäe-ABL]

'Pupils come to us also from the surroundings of Jõhvi, **[from Mäetaguse]**, [from Ahtme] and even [from Sillamäe].'

In Table 24 we provide an overview of the synonymy of spatial cases in this class of TOPOs.

Table 24: Synonymous word-forms in the paradigm of an Estonian TOPO.

relation	set				meaning
	interior		exterior		
Place	INESS	Mäetaguse-*s*	ADESS	Mäetaguse-*l*	'in Mäetaguse'
Goal	ILL	Mäetaguse-*sse*	ALL	Mäetaguse-*le*	'to Mäetaguse'
Source	ELA	Mäetaguse-*st*	ABL	Mäetaguse-*lt*	'from Mäetaguse'

Since the two cases of each row in Table 24 fulfil identical functions in terms of expressing a given spatial relation, we are confronted with overabundance. However, it is not true that the above model holds for each and every TOPO that allows for being inflected in the spatial cases of both sets.

In connection to the issue of illative allomorphy, the case of the name of the village *Jämejala* is especially instructive. This TOPO belongs to the minority of TOPOs which are compatible with both sets of spatial cases. The TOPO contains the final unit -*jala* (< body-part noun *jalg* 'foot'). As a COMM, *jalg* 'foot' reflects the usual pattern of overabundance with the illative which manifests itself in the coexistence of illative-1 *jala-sse* 'into the foot' and illative-2 *jalga* 'into the foot'. The TOPO *Jämejala* has generalised the short illative (= *Jämejalga* 'into Jämejala'). This is striking because the three other TOPOs with final -*jala* which are reported in Erelt (2002) behave like the COMM *jalg* 'foot' in the sense that their paradigms display allomorphy in the cell of the illative, namely
- *Kautjala* → illative-1 *Kautjala-sse* ~ illative-2 *Kautjalga* 'into Kautjala',
- *Laimjala* → illative-1 *Laimjala-sse* ~ illative-2 *Laimjalga* 'into Laimjala', and
- *Mustjala* → illative-1 *Mustjala-sse* ~ illative-2 *Mustjalga* 'into Mustjala'.

---

98 http://www.cs.ioc.ee/~opleht/Arhiiv/99Apr30/koolilood.html

These three TOPOs are distinguished further from *Jämejala* and also from the COMM *jalg* 'foot' by their inability to inflect for the cases of the exterior set. *Jämejala* and *jalg* 'foot' are compatible with the cases of the exterior set as well as with those of the interior set. However, according to Erelt (2002) the choice of exterior or interior set makes no difference semantically in the case of the TOPO *Jämejala* whereas the two sets are not synonymous in the case of the COMM. Accordingly, the following pairs of word-forms of the TOPO should mean exactly the same:
– inessive-adessive: *Jämejala-s* ~ *Jämejala-l* 'in Jämejala',
– illative-allative: *Jämejalga* ~ *Jämejala-le* 'to Jämejala', and
– elative-ablative: *Jämejala-st* ~ *Jämejala-lt* 'from Jämejala'.

On closer inspection, however, this supposed case of free variation invites a new interpretation. Since the late 1960s there is a psychiatric hospital in *Jämejala* which is referred to frequently under the TOPO of the village in our electronic corpus of Estonian. When reference is made to this mental institution, the cases of the interior set are employed. Those of the exterior set are made use of if the village itself serves as the Ground of the spatial situation, cf. (88)–(90).

(88) Estonian [Uralic, Finnic] – Place
    (a) inessive / hospital[99]
        sundraviteenusel   **[Jämejala-s]**   on   inimesed
        enforced_treatment:ADESS   **[Jämejala-INESS]**   be.3SG   already
        juba   sellised
        such:PL   man:PL
        '...for enforced treatment **[in Jämejala]** there are already such persons...'
    (b) adessive / village[100]
        Olen   **[Jämejala-l]**   sündinud
        be:1SG   **[Jämejala-ADESS]**   be_born:PTCPL
        ja   oma   esimesed   aastad   elanud.
        and   POR.RFL   first:PL   year:PL   live:PTCPL
        'I was born **[in Jämejala]** and have passed my first years there.'

---

[99] http://uudised.err.ee/v/eesti/86fa6e77-fee8-49ae-a2f9-8b0408b9daaf
[100] http://www.skala.ajaleht.ee/2462589/kommentaar-jamejalale-moeldes

(89) Estonian [Uralic, Finnic] – Goal
    (a)   illative / hospital[101]
        Mine ravile, enne kui sind **[Jämejalga]**
        go:IMPTV.2SG cure:ALL before when 2SG:PART **[Jämejala.ILL]**
        viiakse.
        take:IMPERS
        'Go and see a doctor before they take you **[into Jämejala]**.'
    (b)   allative / village[102]
        Pärast kooli, kui nad tööle läksid,
        after school when 3PL work:ALL go:PRET:3PL
        tuligi Henri isa siia **[Jämejala-le]**.
        come:PRET:3SG:too Henry.GEN father hither **[Jämejala-ALL]**
        'After school, when they went to work, Henry's father too came here **[to Jämejala]**.'

The formal differentiation is largely uncontroversial with the spatial relations of Place and Goal as shown in Examples (88)–(89): the cases of the interior set refer to the hospital whereas those of the exterior set refer to the village of the same name. The situation is slightly more ambiguous in the case of Source, cf. (90).

(90) Estonian [Uralic, Finnic] – Source
    (a)   elative / hospital[103]
        [Jämejala-st] põgenes ohtlik mõrvar.
        [Jämejala-ELA] escape:PRET.3SG dangerous killer
        'A dangerous killer escaped **[from Järmejala]**.'
    (b)   ablative / village ~ hospital[104]
        [Jämejala-lt] põgenenud tapja andis end
        [Jämejala-ABL] escape:PTCPL killer give:PRET.3SG RFL:PART
        politseile üles.
        police:ALL up
        'The killer who had escaped **[from Jämejala]** gave himself up to the police.'

---

101 http://suusk.blogspot.de/2015/02/pullerits-mis-juhtus-tartu-maratonil.html
102 http://www.ngo.ee/sites/default/files/files/keskkond-ja-kodanikualgatus.pdf
103 http://www.delfi.ee/news/paevauudised/krimi/jamejalast-pogenes-ohtlik-morvar-taiend?id=14426229
104 http://www.ohtuleht.ee/211912/jamejalalt-pogenenud-tapja-andis-end-politseile-ules

The elative in (90a) invites the reading according to which the killer escaped from the hospital where he was kept in custody. As to the interpretation of the ablative in (90b), one could argue that the same situation is referred to, namely the killer's escape from the hospital. However, we understand the ablative to mean that the killer had not only left the hospital but also the village of Jämejala. This interpretation needs to be checked against further empirical evidence. For the time being it suffices to conclude that in the case of *Jämejala*, the two sets of spatial cases are exploited creatively to distinguish the TOPO which identifies the village of Jämejala from the medical institution which is located in the same village. In the latter case, we are no longer dealing with a genuine TOPO but with the metonymic use of a TOPO to label an institution which runs a number of buildings in Jämejala. On this basis, we claim that there is no free variation at all – and that the use of the interior set of spatial cases is not a property of the TOPO *Jämejala* in the first place. The TOPO only tolerates the use of the exterior set of spatial cases. This is a remarkable fact because all other Estonian TOPOs with final *-jala* are compatible exclusively with the interior set of spatial cases (cf. above). Thus, identical morphological conditions may yield different results. It is plausible to assume that the case patterns which are associated with the COMM *haigla* 'hospital' have influenced the differentiation we have observed for *Jämejala*. The COMM *haigla* 'hospital' takes the spatial cases of the interior set to indicate general location as *haigla-s* **'in** the hospital', *haigla-sse* **'into** the hospital', *haigla-st* **'from** the hospital'. However, this explanation cannot be generalised over all cases of (supposed or real) free variation of the two sets of spatial cases of Estonian.[105]

---

[105] A case in point is the name of the village *Pöide* which is reported to combine with the spatial cases of both sets (Erelt 2002). However, at http://www.delfi.ee/newspaevauudised/krimi/paastjad-leidsid-poidelt-kolme-eri-murki?id=20004510&com=1&reg=0&no=0&s=1 we have found an interesting exchange of ideas as to the correct use of the spatial cases with this TOPO: *A: Mäletamist mööda [recte: mööda] vanasti sai mindud [Pöide] ja tuldud [Pöide-st]. B: Täiesti nõus sinuga. Oli jah vanasti minek [Pöide] või [Pöide-sse] ja tuldud ikka [Pöide-st]. C: Ikka lähen [Pöide-le] ja tulen [Pöide-lt]. Te pole kohalikud, seetõttu nii räägitegi, ma ole aga kohalik ;-)* 'A: To my memory people in the old days went **[to Pöide]**$_{\text{illative-2}}$ and came **[from Pöide]**$_{\text{elative}}$. B: I agree with you completely. Yes, in the past there was the walk **[to Pöide]**$_{\text{illative-2}}$ or **[to Pöide]**$_{\text{illative-1}}$ and one always came **[from Pöide]**$_{\text{elative}}$. C: I always go **[to Pöide]**$_{\text{allative}}$ and come **[from Pöide]**$_{\text{ablative}}$. You are no locals, that is why you talk like this, I am from here however ;-)' The practices of insiders (locals) and outsiders (migrants) are compared to each other. The contemporary variation is given a diachronic explanation, i.e., the former preference for the spatial cases of the exterior set is depicted as giving way to the dominance of the interior set which is caused by the influx of people from other parts of Estonia where the local traditions of Pöide are largely unknown. Thus, we are witnessing a process of language change.

Figure 33 meets the expectations in the sense that it looks very much the same as Figure 31 which pictured the situation in Finnish.

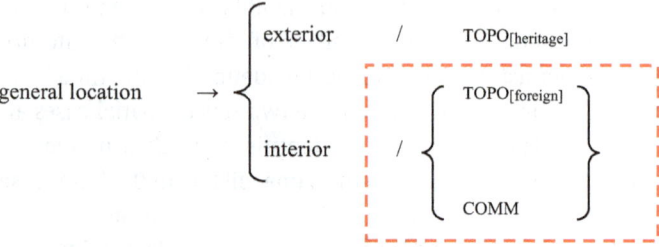

Figure 33: Split rule for general location in Estonian.

Accordingly, the same provisos and limitations as those formulated in connection with Finnish (and Hungarian) also hold for Estonian when it comes to interpreting the simplified account of the complex situation on the basis of Figure 33. The class of TOPOs is divided in two with the inherited TOPOs behaving differently from the bulk of the TOPOs and COMMs.

5.2.4 What the Uralic data tell us

In conclusion, what can be said with reference to the Uralic languages surveyed so far is that all three of them give clear evidence of special morphological behaviour of TOPOs as opposed to ANTHs and COMMs. This special behaviour consists mainly in the blocking of one of the sets of spatial cases which are commonly distinguished with COMMs. With TOPOs, these distinctions are blocked only in the domain of general location. Estonian, Finnish, and Hungarian also agree among each other as to the generalisation of one particular set of spatial cases over all foreign TOPOs. Only for TOPOs which refer to settlements within the (historical) boundaries of the country whose language is at issue is there competition between two options. On the lower levels of the morphological particulars, the languages find different solutions for similar problems. This unsurprising disagreement does not impair the conclusion that TOPOs constitute a class of their own in the morphological systems of Estonian, Finnish, and Hungarian.

However, we emphasise that the separation of TOPOs from ANTHs and COMMs is not as neat as our above statement might suggest. To make our point, we have tacitly assumed that all COMMs form a morphologically homogeneous class. This assumption rests on shaky ground since it is very likely that, for semantic reasons, by no means all COMMs inflect usually for all sets of spatial cases for the purpose of general location. This may be the case with nouns which represent abstract notions, etc. In the absence of a dedicated in-depth study of the compatibility of COMMs with the different sets of spatial cases in the Uralic languages, we have to make do with assuming a continuum on which prototypical TOPOs are distinguished from prototypical COMMs. Less prototypical COMMs (i.e., those which do not lend themselves easily to being inflected in all sets of spatial cases under general location) are situated in the space between the two prototypes, cf. Figure 34.

prototype	less prototypical	prototype
TOPOs	COMMs	
one set less	semantically-based restrictions	all sets

Figure 34: Continuum of prototypicality / Uralic.

5.3 Basque

In contrast to the aforementioned Uralic languages and Lezgian, the special grammar of the TOPOs of Basque does not manifest itself covertly, i.e., in their inability to inflect for the full array of spatial cases in all appropriate syntactic contexts. What makes TOPOs special in Basque is their internal morphological structure which differs from that of ANTHs and COMMs. The decisive criterion is the absence/presence of certain morphemes which mark the semantic class to which a given nouny word belongs. In this sense, we are dealing with an overt phenomenon that distinguishes TOPOs from other word-classes visibly albeit in a peculiar, i.e., unexpected way as we will show in the subsequent paragraphs.

In conformity to Estonian, Finnish, Hungarian, and Lezgian, Basque is characterised by a richly equipped system of morphological cases. In the literature, the size of the Basque case system is said to comprise ten (De Rijk 2008: 31 and 50), twelve (Lafitte 1998: 55), fourteen (Villasante 1972: 18–19), fifteen (Hualde and Ortiz de Urbina 2003: 173), or even seventeen (Bendel 2006: 23) distinct categories. To avoid confusion, we take the most recent prescriptions of Euskara Batua, the unified standard Basque, as our frame of reference since this is representative of the system of fifteen to sixteen cases that is most likely to be used in the modern literary texts which make up our corpus although the systems cannot be considered to be exempt from variation (Hualdea and Ortiz de Urbina 2003: 6, Bendel 2006: xii).

Wherever necessary we also discuss evidence from Basque regional varieties. Unsurprisingly, spatial cases form a major subdivison of the Basque noun paradigm no matter how many categories the case system is assumed to contain.

For COMMs, the declension is divided into four sub-paradigms in correlation to a hierarchy of distinctions. On the highest level, indefinite vs definite are distinguished. Since indefinite NPs are transnumeral per definition, number distinctions are possible exclusively under definiteness. If definiteness applies, nouns distinguish singular and plural. In the plural, it is additionally possible to contrast a general plural from a proximate plural the latter being used to encode the speaker's empathy. The proximate plural is considered a nonstandard feature in Hualde and Ortiz de Urbina (2003: 122) whereas it is treated on a par with the general plural in Bendel (2006: 17–18).[106] We exemplify the Basque situation with the full paradigm of the COMM *leku* 'place' as given in Hualde and Ortiz de Urbina (2003: 173). Note that this noun has a vocalic stem which requires the insertion of the euphonic *-r-* in combination with vowel-initial suffixes (Hualde and Ortiz de Urbina 2003: 175). The case labels are those which our source employs (other descriptions of Basque have different terms for several of the cases). We refrain from burdening the presentation of the paradigm with approximate translations into English. The spatial cases will be scrutinised below so that there is ample opportunity to specify their meaning. Boldface marks out the morpheme which indicates inanimacy and which is crucial in the discussion to follow.

- *leku* 'place' →
    - indefinite: absolutive *leku*, ergative *leku-k*, dative *leku-r-i*, genitive *leku-r-en*, benefactive *leku-r-entzat*, comitative *leku-r-ekin*, instrumental *leku-z*, partitive *leku-rik*, prolative *leku-tzat*, locative *leku-**ta**-n*, ablative *leku-**ta**-tik*, allative *leku-**ta**-ra*, directive *leku-**ta**-rantz*, terminative *leku-**ta**-raino*, relative *leku-**ta**-ko*
    - definite:
        - singular: absolutive *leku-a*, ergative *leku-a-k*, dative *leku-a-r-i*, genitive *leku-a-r-en*, benefactive *leku-a-r-entzat*, comitative *leku-a-r-ekin*, instrumental *leku-a-z*, locative *leku-a-n*, ablative *leku-tik*, allative *leku-ra*, directive *leku-rantz*, terminative *leku-raino*, relative *leku-ko*
        - plural:
            - general: absolutive *leku-a-k*, ergative *leku-e-k*, dative *leku-e-i*, genitive *leku-e-n*, benefactive *leku-e-ntzat*, comitative *leku-e-kin*, instru-

---

106 It is also not included in De Rijk's (2008) account of Standard Basque. Villasante (1972: 105) calls the proximate plural "[u]n fenómeno también universal de la lengua" (Our translation: "also a pervasive phenomenon in the language").

mental *leku-e-z*, locative *leku-e-**ta**-n*, ablative *leku-e-**ta**-tik*, allative *leku-e-**ta**-ra*, directive *leku-e-**ta**-rantz*, terminative *leku-e-**ta**-raino*, relative *leku-e-**ta**-ko*
- proximate: absolutive *leku-o-k*, ergative *leku-o-k*, dative *leku-o-i*, genitive *leku-o-n*, benefactive *leku-o-ntzat*, comitative *leku-o-kin*, instrumental *leku-o-z*, locative *leku-o-**ta**-n*, ablative *leku-o-**ta**-tik*, allative *leku-o-**ta**-ra*, directive *leku-o-**ta**-rantz*, terminative *leku-o-**ta**-raino*, relative *leku-o-**ta**-ko*

The partitive and the prolative are banned from contexts with the feature [+definite] and thus show up only in the paradigm of the indefinite noun. For the purpose at hand, the number distinctions as such are only of secondary relevance if at all since TOPOS usually do not participate in the singular-plural opposition. What is of primary relevance, however, is the different morphological structure of the various sub-classes of nouns which comes to the fore especially in the word-forms of the spatial cases.

In point of fact, three classes are distinguished formally, namely inanimate COMMS, animate nouns, and TOPOS. As the above paradigm of *leku* 'place' shows, except in the definite singular, spatial cases and the so-called relative case of COMMS are represented by word-forms which involve the non-final suffix *-ta-*. This morpheme can be taken to be the marker of inanimacy. In contrast, animacy is marked obligatorily on definite and indefinite animate nouns in those word-forms which host the suffixes of the spatial cases. The marker of animacy is *-gan-* which is usually attached to a stem which is identical to the genitive in *-en* (but cf. below).[107] In the case of TOPOS, however, neither of these markers is ever employed. What is more, there is no class marker at all in the word-forms of TOPOS. Simplifying we

---

107 Anderson (2007: 304) interprets the Basque facts differently insofar as he assumes that "in Basque personal names do not inflect for the locative cases found with nouns and place names. [. . .] [T]he (directive) allative suffix *-rat* [is] attached respectively to a common noun and a place name [. . .]. In order to construct an 'allative' for a name [. . .] the allative of a postposition is deployed (*ganat*), and the name is in the possessive genitive; the direct suffixing of the allative in [*San Martinerat* 'to(wards) St. Martin (town)'] tells us that this is the name of a locality and not a person. This is but one indication that personal names are more central to the class than even settlement names – unsurprisingly, given the anthropocentricity of language." Thus, Anderson postulates multi-word constructions – postpositional phrases, that is – for all spatial cases of ANTHS. What he overlooks is that ANTHS behave like animate COMMS morphologically, i.e., there is no privilege of this class of names in the first place (see below). Moreover, the obligatory marking of inanimacy with COMMS is not acknowledged so that a crucial difference is glossed over that distinguishes inanimate COMMS from TOPOS.

can assume a pattern of distribution as shown in Figure 35 (which will be refined further below).

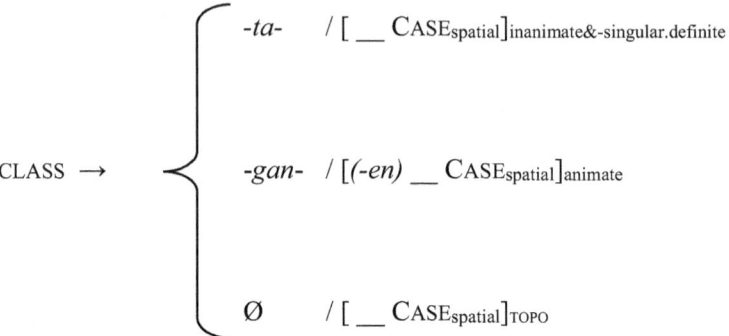

Figure 35: Distribution of class-markers with Basque nouns.

Superficially, this looks like a very straightforward formal distinction of the three classes. However, the situation is made less clear-cut because of the absence of the class marker on definite animate nouns in the singular. Since this is tantamount to zero-marking of the class-membership, there is the possibility that inanimate nouns in the definite singular are formally indistinguishable from TOPOs because the latter display no class-marker either.

However, Table 25 is indicative of an even more intricate constellation of facts. In Table 25, we survey the word-forms of three grammatical cases – absolutive, ergative, and relative – and those of the spatial cases – locative, ablative, allative, directive, and terminative – of two COMMs and a TOPO, namely inanimate *azal* 'skin', animate *mutil* 'boy', and TOPO[108] *Irun* (Hualde and Ortiz de Urbina 2003: 174–177) in the non-plural. All three of these Basque words have consonant-final stems and thus fulfil identical phonological criteria. Grey shading marks cells which host word-forms reflecting identical internal morphological composition for all three nouns. Single underlining identifies those word-forms which display an identical morphological structure for the inanimate noun and the TOPO. Double underlining is used to highlight those word-forms which are morphologically similar between the two COMMs. In contrast, boldface marks those word-forms which are built according to a pattern that is unique to the TOPO.

---

[108] In this section we focus on the names of settlements. Other macrotoponyms such as the names of countries seem to behave largely in accordance to the names of settlements. For a number of peculiarities of sub-classes of TOPOs, the reader is referred to De Rijk (2008: 57–59).

Table 25: Selected cases of the paradigms of Basque inanimate nouns, animate nouns, and TOPOs.

case	inanimate noun		animate noun		TOPO
	indefinite	definite	indefinite	definite	
ABS	*azal*	*azal-a*	*mutil*	*mutil-a*	*Irun*
ERG	*azal-e-k*	*azal-a-k*	*mutil-e-k*	*mutil-a-k*	*Irun-e-k*
LOC	*azal-e-ta-n*	*azal-e-a-n*	*mutil-en-gan*	*mutil-a(-r-en)-gan*	**Irun-e-n**
ABL	*azal-e-ta-tik*	*azal-e-tik*	*mutil-en-gan-dik*	*mutil-a(-r-en)-gan-dik*	**Irun-dik** ~ *Irun-e-tik*
ALL	*azal-e-ta-ra*	*azal-e-ra*	*mutil-en-gan-a*	*mutil-a(-r-en)-gan-a*	**Irun-a** ~ *Irun-e-ra*
DIR	*azal-e-ta-rantz*	*azal-e-rantz*	*mutil-en-gan-atz*	*mutil-a(-r-en)-gan-atz*	*Irun-e-rantz*
TERM	*azal-e-ta-raino*	*azal-e-raino*	*mutil-en-gan-aino*	*mutil-a(-r-en)-gan-aino*	*Irun-e-raino*
REL	*azal-e-ta-ko*	*azal-e-ko*	---	---	**Irun-go** ~ *Irun-e-ko*

What strikes the eye immediately is the empty cell of the relative case with the animate noun. This gap is general for all animate nouns (Villasante 1972: 19).[109] Moreover, similarities of word-forms in all three of the paradigms are restricted to the absolutive and ergative. In these cases, the TOPO follows the same morphotactic pattern as the COMMs in the indefinite whereas inanimate and animate nouns resemble each other morphologically in the same cases of the definite sub-paradigm.

As to the spatial cases, the resemblance is less clear. First of all, the animate noun shares the morphological structure of its spatial cases with none of the two other words. Secondly, wherever the TOPO and the inanimate noun display correspondence as to the morphological make-up of the word-forms of the spatial cases, these similarities involve the definite sub-paradigm of the COMM. Thus, there is a split between the grammatical cases and the spatial cases, in terms of which of the sub-paradigms provides the word-forms which fit those of the TOPO. Thirdly, the parallelism of the morphology of the spatial cases of the inanimate noun and the TOPO is complete only for the directive and the terminative. For the ablative and the allative, however, the TOPO attests to overabundance since each of the cases can be expressed by two distinct synonymous word-forms. Only one of the two alternatives has a direct morphological equivalent in the paradigm of the inanimate noun – the competing word-forms being unique to the TOPO. This pattern also

---

109 De Rijk (2008: 89–91) argues vehemently against considering *-ko* a case affix in the first place.

holds for the relative case. Note that overabundance as such is a property which none of the other sample words shares with the TOPO. Finally, there is the locative case whose morphological representation on the COMM is different from that of the same case on the TOPO. In sum, TOPOs display a declension which is a patchwork of the patterns of the indefinite and the definite sub-paradigms of inanimate COMMs in combination with a pattern that is exclusive to TOPOs. No other nouny word-class shares exactly this combination of structural properties so that it is legitimate to judge TOPOs to constitute a morphologically-defined class of their own.

What exactly distinguishes the word-forms of TOPOs and definite inanimate nouns formally needs to be determined on the level of morphophonology. Definite inanimate nouns insert an epenthetic vowel /e/ between a stem-final consonant and a consonant-initial affix. This applies to the ablative (whose exponent is -*tik*) and the relative (whose exponent is -*ko*). Thus, the COMM *ibar* 'lowlands' is *ibarr-e-tik*[110] 'from the lowlands' in the ablative and *ibarr-e-ko* 'of the lowlands' in the relative (Villasante 1972: 99). Hualde and Ortiz de Urbina (2003: 177) claim, however, that

> [i]n the ablative and relational of proper names ending in a consonant, -*e*-epenthesis is optional. When the name ends in -*n* or -*l*, forms without epenthesis and with voicing of the suffix-initial consonant are preferred in the standard language, but both alternatives are found in local dialects.

This means that the class-membership of a nouny item determines whether or not the requirements of morphophonology can be overruled. This becomes even more evident when the morphophonology of animate nouns is taken into account, too. Their class marker -*gan* ends in a nasal consonant to which the case suffixes are attached directly without any kind of epenthesis. Accordingly, the noun *seme* 'son' not only displays
- the ablative *seme-a-r-en-gan-dik* 'from the son' but also
- the allative *seme-a-r-en-gan-a* 'to the son',
- the directive *seme-a-r-en-gan-antz* 'towards the son', and
- the terminative *seme-a-r-en-gan-aino* 'as far as the son'.

Thus, the initial segment of the case affix is morphophonologically irrelevant as none of the suffixes triggers the insertion of additional phonological material. Nevertheless, some of these forms need to be discussed further below. The above word-forms of the animate COMM share the following pattern of morphological structure:
- {LEX}-{DEF}-{LIG}-{GEN}-{CLASS.ANIM}-{CASE}

---

110 Taps and trills are neutralised word-finally as singleton /r/. The underlying trill surfaces if it is followed by a vowel under affixation (Hualde and Ortiz de Urbina 2003: 30–31).

Definite inanimate nouns and TOPOs do not copy this template. The corresponding word-forms of the former show that epenthesis is compulsory:
- ablative *azal-e-tik* 'from the skin',
- allative *azal-e-ra* 'to the skin',
- directive *azal-e-rantz* 'towards the skin',
- terminative *azal-e-raino* 'as far as the skin',
- relative *azal-e-ko* 'of the skin'.

The insertion of *-e-* in the above examples is explained by Hualde and Ortiz de Urbina (2003: 174) as a strategy to "break consonant sequences". This explanation is convincing if we accept that the case affixes of the allative, directive, and terminative have a basic form with an initial rhotic consonant (i.e., *-ra, -rantz,* and *-raino*). Like Lafitte (1998: 55) and many others, De Rijk (2008: 49) takes *-ra(-)* to be the basic form of the allative suffix and of derivatives thereof so that the absence of the initial rhotic in some contexts must be explained as a case of consonant deletion (De Rijk 2008: 57). This applies for instance to the animate nouns above which have *-a, -antz,* and *-aino* attached to the consonant-final animacy marker. As will be shown shortly, TOPOs also allow for the allative in *-a* on stems with final consonant.

In terms of economy, De Rijk's (2008) analysis is no improvement since the complexity of the rules of the differential morphophonological behaviour of TOPOs, ANTHs, and COMMs is not reduced. Alternatively, one could consider the alternation of word-forms with and without /r/ on the boundary between affix and stem to be an instance of excrescence and its absence. However, Hualde and Ortiz de Urbina (2003: 175) assume that the /r/ is epenthetic only in those cases where stems with final vowels and those with final consonants behave differently (e.g. with the genitive in the indefinite *leku-r-en* vs *azal-en* (and not **azal-e-r-en*)). Independently of this differentiation of epenthetic /r/ and affix-initial "underlying" rhotic, the presence of the epenthetic /e/ is not explained conclusively by way of invoking the avoidance of consonant sequences.[111] The morphophonological behaviour of the animate nouns casts additional doubt on the purely morphophonological reasoning of the reference grammar. Moreover, TOPOs also support the idea that the above-mentioned interpretation is insufficient. The name of the city of Eibar reflects overabundance in three of the cells of the paradigm (Hualde and Ortiz de Urbina 2003: 177):
- ablative *Eibarr-e-tik ~ Eibar-tik* 'from Eibar',
- allative *Eibarr-e-ra ~ Eibarr-a* 'to Eibar' (= allomorphy of *-ra ~ -a*),

---

[111] Hualde and Ortiz de Urbina (2003: 60) state that "[i]t should be noticed that epenthesis takes place even when the resulting consonant sequence would still be well formed without it." This stating of the facts does not explain the variation between the different nouny word-classes discussed in this paragraph.

- directive *Eibarr-e-rantz* 'towards Eibar',
- terminative *Eibarr-e-raino* 'as far as Eibar',
- relative *Eibarr-e-ko* ~ *Eibar-ko* 'of Eibar'.

In the directive and the terminative, the word-forms of the TOPOs conform to the patterns that are familiar from the inflection of definite inanimate nouns, i.e., the consonant-final stem is followed by an epenthetic segment /e/ to which the suffixes *-rantz* and *-raino* are attached. The same pattern is also possible with the word-forms of the ablative, allative, and relative. However, in contrast to the former cases, there are also alternative word-forms for each of the three last-mentioned categories. In these alternative word-forms, no excrescent segment shows up since the affixes combine directly with the stem. The comparison of the attested morphological patterns of TOPOs and definite inanimate nouns yields the co-existence of two options, one of which is the monopoly of TOPOs:

- {LEX}-{CASE} / TOPO$_{\text{ablative/allative/relative}}$
- {LEX}-{LIG}-{CASE} / TOPO/definite inanimate$_{\text{dynamic\_spatial\_case/relative}}$

The division of the nouny word-classes into three is especially visible in the formal expression of the locative. As to animate nouns, there is no overt exponent of the locative. As a matter of fact, the marker of animacy *-gan* is followed by case affixes in all spatial cases and the relative with the exception of the locative. It might be argued that in a word form like *gizon-a-r-en-gan* 'in the man' the locative either is expressed by a zero-morpheme (= *gizon-a-r-en-gan-Ø*) or lacks any kind of representation since its function is inherent to the word-form which occupies the appropriate cell of the paradigm (= [*gizon-a-r-en-gan*]$_{\text{locative}}$).[112] Independent of the correct analysis, one thing is clear, namely that the locative of animate nouns is different from the locative of inanimate nouns and TOPOs. There is thus good reason to assume that we are dealing with different inflectional classes (although the morphological class differences correlate also with semantic differences) (Corbett 2007: 30).

Definite inanimate nouns and TOPOs differ among each other on the same parameter, i.e., the formation of the word-forms of the locative follows different principles. Once more morphophonology is crucial. The locative affix is monoconsonantal *-n*. This affix can be attached directly to a host that ends in a vowel such as *mendi* 'mountain' → locative *mendi-ta-n* 'on a mountain' and *Bilbo* 'Bilbao' → locative *Bilbo-n* 'in Bilbao'. If the preceding segment is a consonant, the epenthetic /e/ is inserted as in *Madril* 'Madrid' → locative *Madril-e-n* 'in Madrid'. Worthy of

---

[112] De Rijk (2008: 60) argues that *-gan* is the locative morpheme in the first place.

note is the locative of the definite inanimate noun. With vowel-final stems, we find *mendi-an* 'on the mountain' without the expected insertion of /r/ into the vowel sequence. Consonant-final stems are accompanied by the epenthetic /e/ in combination with the locative marker although the latter is vowel-initial (as e.g. *azal-e-an* 'on the skin'). Hualde and Ortiz de Urbina (2003: 175) argue that

> [t]he reason for this unexpected allomorphy must be that at some historical point this [i.e., the locative marker] was a consonant initial suffix.

The descriptive grammarians of Basque assume that the initial consonant was lost in the course of time but its effects on the surrounding morphological units have remained in place. However, the supposed locative marker *-an* is decomposed into the definiteness marker *-a-* and the locative suffix *-n*, the gravity of the above problem can be reduced at least to some extent. The definiteness marker *-a* attaches to vowel-final stems generally without triggering the insertion of /r/. The definiteness marker is followed by the case affixes, i.e., it is located relatively near to the stem. It is therefore only logical that *-a-n*$_{\text{locative}}$ is a sequence of two morphemes on a par with say, *-a-k*$_{\text{ergative}}$, *-a-r-en*$_{\text{genitive}}$, *-a-z*$_{\text{instrumental}}$, etc. In these parallel cases no epenthetic /e/ is inserted either with vocalic stems (e.g. *leku-a-z* 'with the place') or consonant stems (e.g. *azal-a-z* 'with the skin'). This means that except in the locative there is no trace of a supposed erstwhile affix-initial consonant which should be there if the *-a*$_{\text{definite}}$ is the representative of the same category in all of the word-forms discussed above. A possible solution for this fix is analogy. The epenthetic /e/ is present in all forms of the spatial cases and the relative in the indefinite sub-paradigm of consonantal stems because the inanimate class marker *-ta* requires its presence. Moreover, epenthesis affects all spatial cases and the relative in the definite sub-paradigm. This quasi-ubiquitous epenthetic /e/[113] may have spread to the locative in a process of paradigmatic homogenisation which aimed at giving to the word-forms of the spatial cases a common morphological pattern to distinguish them from the non-spatial cases of the same paradigm. That the vowel in *-a-n* is a morphological unit of its own with the function of encoding definiteness can also be gathered from the locative of TOPOS. Their inherent definiteness blocks the affixation of the definiteness marker. Accordingly, only the bare locative marker is attached to the stem. If the latter ends in a consonant epenthesis applies

---

113 We do not consider the *-e-* of the general plural of COMMS to be epenthetic because
   (a) it does not trigger /r/-insertion on vowel stems (as, e.g., locative general plural *leku-e-ta-n* 'in the places') and
   (b) it alternates regularly with the *-o-* of the proximate plural (as, e.g., instrumental general plural *leku-e-z* 'with the places' vs instrumental proximate plural *leku-o-z* 'with the places (of ours)').

as in *Paris* 'Paris' → *Paris-e-n* 'in Paris'. TOPOs are morphologically different from COMMs – be they animate or inanimate – because in the morphological template of TOPOs there is no slot for the definiteness marker and a class marker. Since two morphological slots are missing from the morphotactics of TOPOs, this word-class boasts a structure which is considerably simpler than that of the COMMs.

In their grammar of Basque, Hualde and Ortiz de Urbina (2003: 177) make the following observation:

> Proper nouns or names take the same endings as indefinite noun phrases in the nonlocal cases; e.g. ergative: *Peru-k, Jon-ek*; dative: *Peru-ri, Jon-i*. In the local cases and with the relational, the suffix is added directly without the infix *-ta-*.

Since *Jon* is a male ANTH and the authors talk sweepingly of names in general, the reader is led to assume that all kinds of names behave the same in Basque. The phenomena mentioned in the above quote in relation to the spatial cases and the relative are exemplified in the reference grammar exclusively with TOPOs. On closer inspection, it is revealed, however, that ANTHs and TOPOs cannot be treated as members of the same class. ANTHs refer to animate beings and thus are morphologically similar to animate nouns in the sense that ANTHs like any other animate noun host the animate class marker in the spatial cases (Villasante 1972: 61–62). The pair of sentences in (91) shows that ANTHs bear the animacy marker in the spatial cases and thus resemble animate COMMs whereas TOPOs do not host an overt class marker. Like animate COMMs, ANTHs take the /r/-less allomorphs of the dynamic spatial cases.

(91) Basque [isolate]
    (a) TOPO {HP I Basque, 61}
        [Londres-e-ra] joan eta eskolarako gauza guztiak erosi
        [London-LIG-ALL] go and school:DEF:REL thing all:DEF:PL buy
        'Going **[to London]** and buying all things for the school...'
    (b) ANTH {HP I Basque, 239}
        [Hagrid-en-gan-a] joan behar dugu oraintxe bertan.
        [Hagrid-GEN-ANIM-ALL] go necessity have.1PL.ERG at_once there
        'We must go **[to Hagrid]** immediately.'

On this basis, we can conclude the section on Basque with a summary of our findings in Table 26. Table 26 shows that on one parameter TOPOs are further removed from ANTHs than they are removed from COMMs. The highest number of shared features is discernible with animate nouns and ANTHs. Grey shading marks shared features which involve animate nouns. Boldface highlights those cases in which the shared features involve inanimate COMMs but no animate nouns.

Table 26: Continuum of nouny word-classes in Basque.

criterion	animate nouns	ANTHS	TOPOS	inanimate nouns	
				DEF	INDEF
class marker	yes	yes	no	no	yes
epenthesis	blocked	blocked	optional	yes	yes
r- → ∅ / [spatial]	yes	yes	(allative)	no	no
special locative	yes	yes	yes	no	no

In conclusion, the above discussion is suggestive of the special status of TOPOS in Basque which constitute a class of their own which is characterised by a combination of properties which is attested with no other nouny word-class of the same language. As in Lezgian and the Uralic languages, the special behaviour of TOPOS surfaces in the context of the expression of spatial relations.

5.4 Antiquity

In this section, we present data from Ancient Greek and Latin which testify to a combination of properly morphological facts and aspects which transcend the word boundaries. On the one hand, TOPOS can be shown to preserve inflections that have disappeared from the paradigms of COMMS. At the same time, the characteristic morphological traits of TOPOS interact with the grammar of prepositions in both Ancient Greek and Latin. Luraghi (2009) addresses this issue from the point of view of the diachronic development of case systems from Proto-Indo-European to the two classical languages. In her study, the role played by TOPOS is emphasised repeatedly (Luraghi 2009: 289–290 (for Ancient Greek), 297–300 (for Latin)) although the author does not focus specifically on this word-class. It makes sense therefore to scrutinise the givens anew. We proceed chronologically, i.e., we start by looking at the Ancient Greek data and then turn our attention to those of Latin.

5.4.1 Ancient Greek

If we discount the vocative, the case system of Ancient Greek comprises four categories, namely nominative, accusative, dative, and genitive. COMMS, ANTHS, and TOPOS are declinable generally. As expected, names normally come only in one number. Beside those TOPOS which are always singular as, e.g., *Spártē* 'Sparta', there is also a number of TOPOS that are usually inflected according to the plural sub-paradigm of COMMS as, e.g., *Athḗnai* 'Athens' (for which there is also the more rarely

used singular *Athḗnē*). We illustrate the general picture with the paradigms of four representative items – which means that we gloss over a wealth of morphological peculiarities of inflectional classes and individual nouns since these facts are not of primary relevance for the issue at hand (Bornemann and Risch 1978: 26–49):
- COMM: *lógos* 'word, speech' →
    - singular: nominative *lóg-os*, accusative *lóg-on*, genitive *lóg-ou*, dative *lóg-ōĭ*
    - plural: nominative *lóg-oi*, accusative *lóg-ous*, genitive *lóg-ōn*, dative *lóg-ois*
- ANTH: *Diogénēs* 'Diogenes' → nominative *Diogén-ēs*, accusative *Diogén-ē*, genitive *Diogén-ous*, dative *Diogén-ei*
- TOPO:
    - singular: *Spártē* 'Sparta' → nominative *Spárt-ē*, accusative *Spárt-ēn*, genitive *Spart-ēs*, dative *Spart-ēĭ*
    - plural: *Athḗnai* 'Athens' → nominative *Athḗn-ai*, accusative *Athḗn-as*, genitive *Athḗn-ōn*, dative *Athḗn-ais*

In addition to their functions as cases of the direct and indirect object of transitive and ditransitive verbs, the accusative and dative are employed also for the purpose of encoding spatial relations. To a more limited extent, this is also true of the genitive.

The dative is used in static contexts, i.e., in connection with the spatial relation of Place (Bornemann and Risch 1978: 196). This function is inherited from the erstwhile locative case which has merged with the dative very early on in the language history of Ancient Greek – presumably long before the first written documents saw the light of day (Luraghi 2009: 288). TOPOS and COMMS display different particulars as to the expression of the spatial relation of Place. For COMMS, the use of the preposition *en* (~ *enì* ~ *ein*) 'in' in combination with the noun in the dative is almost compulsory[114] whereas, in the case of TOPOS, the very same preposition is optional, i.e., the bare TOPO in the dative is sufficient to express Place (Luraghi 2009: 289), cf. (92)–(93).

---

[114] We cannot say much as to the situation in the domain of Ancient Greek ANTHS. The sole example of a spatial relation being expressed with an ANTH as Ground in our corpus text is {Odu 22, 95–96} *Tēlémaxos d' apórouse, lipṑn dolixoskion égxos autoũ [en Anphinóm-ōĭ]$_{PP}$.* 'Telemach, however, jumping back let the shadow-offering spear stuck **[in Anphinomos]**.' with the ANTH in the dative governed by the preposition *en* 'in'. On this insufficient basis, it could be hypothesised that ANTHS behaved largely like COMMS.

(92) Ancient Greek [Indo-European, Greek] – COMM / Place {Odu[115] 10, 221–222}
oudé   toi   'huías   zṓein   **[en megár-oisin]**_{PP}   eásomen
not_even well   son:ACC.PL live:INF **[in hall-DAT.PL]**_{PLACE} let:FUT:1PL
'...and we won't let [your] sons live **[in the halls]**...'

(93) Ancient Greek [Indo-European, Greek] – TOPO / Place
 (a) PP {Odu 4, 6}
    [en Troí-ēï]_{PP}   gàr   prōton   hypésxeto
    [in Troja-DAT]   because   first   entertain:AOR.MED.3SG
    'Since he agreed at first **[in Troia]**...'
 (b) bare noun / dative {Odu 6, 162}
    [Dél-ōĭ]   dé   pote   toĩon   Apóllōnos   parà   bōmō̃ĭ [...]
    [Delos-DAT]   now   once   so   Apollo:GEN at   altar:DAT [...]
    enóēsa
    AUGM:perceive:AOR.1SG
    'I once saw an altar dedicated to Apollo **[in Delos]**...'

The bare dative used spatially "s'observe surtout dans des expressions de sens assez général"[116] (Chantraine 1963: 78), among which body-part terms and fixed formulaic expressions stand out (Chantraine 1963: 78–79). According to Luraghi (2009: 289) there are also mostly adverbialised COMMs whose dative reflects the erstwhile locative as, e.g.,
– pónt-ōĭ 'in [the] sea' ← pónt-os 'sea' and trápez-ēi 'at (the) table' ← trápez-a 'table' (Luraghi 2009: 289)).

Luraghi (2009: 290–291) argues that a subset of those NPs for which the use of the above preposition en (~ enì ~ ein) 'in' is not obligatory for the expression of the spatial relation of Place also allows for the so-called "allative accusative", i.e., the morphological accusative encodes the spatial relation of Goal (Bornemann and Risch 1978: 183). Similarly to the previous case, the accusative may be optionally (but nevertheless predominantly) accompanied by a preposition (= es ~ eis 'to' (< en + -s)) or (much more rarely) stand alone, cf. (94)–(95).

---

[115] For the purpose of this study, we look exclusively at Ancient Greek data from the Homeric era. More specifically, we rely empirically on the evidence found in the classical text of the Odusseias (dating back to the eighth pre-Christian century). Thus, our hypotheses still have to stand a future test conducted according to philological criteria of the Classics – a project which is far beyond the limits of our cross-linguistically-oriented investigation.
[116] Our translation: "is observed especially in expressions of rather general meaning".

(94) Ancient Greek [Indo-European, Greek] – COMM / Goal {Odu 17, 6}
Att' ễ toi mèn egṑn eĩm' ***[es pól-in]***
father really well surely 1SG go:1SG **[to town-ACC]**
'Well, father, I will be going **[to town]** now...'

(95) Ancient Greek [Indo-European, Greek] – TOPO / Goal
  (a) PP {Odu 15, 1–2}
      Hễ d' ***[eis eurúxor-on Lakedaímon-a]***_{PP}
      well however **[to wide-ACC Lakedaimon-ACC]**
      Pallàs Athḗnē ṓïxet'
      Pallas Athene move:IMPERF.3SG
      'Well then, Pallas Athene went **[to the extended Lakedaimon]**...'
  (b) bare noun/accusative {Odu 8, 362}
      Hễ d' ára ***[Kúpr-on]*** *híkane philommeidḗs*
      well however then **[Cyprus-ACC]** come:IMPERF.3SG smiling
      Aphrodítē
      Aphrodite
      'Well, then the smiling Aphrodite went **[to Cyprus]**...'

Luraghi (2009: 290) argues that the allative accusative is possible mostly with COMMs which indicate Goals. The pair of sentential examples in (96) is indicative of the optional character of the prepositionless encoding of the Goal relation with COMMs.

(96) Ancient Greek [Indo-European, Greek] – COMM / Goal
  (a) PP {Odu 12, 9}
      dề tót' egṑn hetárous proḯen
      now then 1SG comrade:ACC.PL send_forth:IMPERF.1SG
      [es dṓmata Kírkēs]_{PP}
      [to palace:ACC.PL Circe:GEN]
      'I then sent forth the comrades **[to Circe's palace]**...'
  (b) bare noun/accusative {Odu 6, 296} (= Luraghi [2009: 291])
      hikṓmetha ***[dṓmata patrós]***
      go:SUBJ.AOR.1PL **[palace:ACC.PL father:GEN]**
      '...may we go **[to (my) father's palace]**.'

Those COMMs which can express Goal without accompanying preposition constitute the relatively restricted class of TOPO-Ns, i.e., nouns which refer to entities which very commonly function as Grounds in spatial situations. The average COMM, however, does not seem to have the same options. Similarly, bare TOPOs in the accusative are attested only relatively infrequently. More generally, the type and token frequency of the preposition *es ~ eis* 'to' in combination with the allative accusative increased considerably in post-Homeric times.

As to the expression of the dynamic spatial relation of Source, Luraghi (2009: 293) assumes – in agreement with Bornemann and Risch (1978: 189–190) – that

> the possibility of the genitive to denote source is dependent on the verb, while the possibility for the dative to denote location and for the accusative to denote direction is rather dependent on lexical features of the NPs involved. Besides, especially in the case of the dative, independence of the locative meaning from the verb is also shown by the fact that dative NPs with spatial referents can have locative meaning also when they function as adverbials. This never holds for the ablatival genitive: genitive NPs which are syntactically adverbials never express source.

There is thus an asymmetry in the system of spatially employed cases of Ancient Greek. TOPOs do not seem to be responsible for the use of the genitive in ablative function because the semantics of the NP which serves as Ground are irrelevant in the first place. In (97a), the finite verb hosts the prefix *eks-* 'from; out of' which triggers the genitive on the TOPO *Troia*. In (97b), it is the cognate preposition *ek* 'from; out of' which has the same effect on its complement which happens to be the same TOPO.

(97) Ancient Greek [Indo-European, Greek] – TOPO
    (a) bare noun / genitive {Odu 5, 39}
        hós   àn   oudè pote   *[Troí-ēs]*   eksérat'
        REL.M.SG probably NEG ever **[Troia-GEN]** pull_out:SUBJ.3SG
        Odusseús
        Odysseus
        '…which Odysseus would probably never take **[from Troia]**.'
    (b) PP {Odu 18, 259–260}
        ỗ   gúnai,   ou   gàr   oíō   eüknḗmidas
        oh   woman:VOC NEG then think:1SG well_armoured:ACC.PL
        Axaioùs   *[ek   Troí-ēs]*$_{PP}$ eũ   pántas
        Achaian:ACC.PL **[from Troia-GEN]**$_{PP}$ well all:ACC.PL
        apḗmonas   *aponéesthai.*
        unharmed:ACC.PL return:INF:FUT
        'Oh woman, I do not think that the well-armoured Achainas will all return unharmed **[from Troia]**.'

Luraghi (2009: 292–294) demonstrates that the genitive is different from the dative and the accusative in terms of its interpretation in the context of spatial relations. Not only is the ablative reading of the genitive dependent on the verb from which the NP is dependent but there are also other usages of the morphological genitive which invokes readings which are incompatible with the spatial relation of Source as, e.g., the partitive function of the genitive which is responsible for the possibility of NPs in the genitive to express Place as in (98).

(98) Ancient Greek [Indo-European, Greek] – TOPO / Source {Odu 3, 251} (= Luraghi [2009: 293])
    ẽ   ouk **[Árge-os]** ẽen
    PTCL NEG **[Argos-GEN]** be:IMPERF.3SG
    'Was he not **[in Argos]**?'

In a footnote, Luraghi (2009: 293, footnote 17) explains that

> [t]he partitive genitive in location expressions indicated special features regarding the internal structure of the landmark, i.e., that the landmark was conceived of as multiplex discontinuous.

We do not intend to inquire further into this issue. What is important for our purpose is the apparently indistinctive behaviour of TOPOs and COMMs in the domain of the spatial relation of Source. However, this impression does not capture the whole story.

The two dynamic spatial relations pose additional problems which are mentioned only in passing in our main reference. According to Luraghi (2009: 289, footnote 13)

> [c]ity names and some other toponyms occur in direction expressions with the prepositionless accusative mostly accompanied by the directive suffix -*de*, a particle that was productively used only in Homeric Greek,

whereas bare COMMs with "spatial reference" (Luraghi 2009: 289) usually occur without this particle if used in the directional accusative. Bornemann and Risch (1978: 183) argue that, in combination with TOPOs the particle =*de* replaces the prepositions. What this means is that preposition and cliticised particle were incompatible with each other and thus could not co-occur in the same NP. Moreover, the clitic seems to be banned from heavy NPs, i.e., from nouns which are accompanied by attributes such as adjectives. What is not mentioned in Luraghi (2009) is the existence of a second clitic =*then* which is employed also only on singleton nouns (i.e., in light NPs) which are involved in the spatial relation of Source. In (99) we illustrate the use of these clitics on TOPOs.

(99) Ancient Greek [Indo-European, Greek]
   (a) Goal {Odu 3, 266–267}
   Atreḯdēs   *[Troí-ēn=de]*   *kiṑn*
   Atreos:GEN   [Troia-ACC=to]   move:PTCPL
   '…when Agamemnon [= (son) of Atreos] moved [to Troia]…'
   (b) Source {Odu 11, 160}
   ẽ   *nũn*   *dḕ*   *[Troíē=then]*   *alṓmenos*
   surely   then   now   [Troia=from]   wander_about:PTCPL.M
   entháď   *híkaneis*
   hither   come:2SG
   'You surely come hither now wandering about [from Troia]…'

As a matter of fact, the above clitics form part of a genuine paradigm of spatial categories which is typical of certain sets of adverbs, pronouns, TOPO-Ns, and last but not least of TOPOs. The basic facts are captured by Table 27 (based on Bornemann and Risch (1978: 58 and 68) and Gemoll (1979: 2, 448, and 533)). Clitics are marked out in bold.

Table 27: Paradigms involving spatial clitics in Ancient Greek.

category	Place		Goal	Source	meaning
adverb	*állo=**thi***		*állo=**se***	*állo=**then***	elsewhere
adverb	*auto=**thi***		*autó=**se***	*autó=**then***	there
TOPO-N	(*oík-oi*)	*oíko=**thi***	*oíka=**de***	*oíko=**then***	home, house
TOPO	(*Megar-oĩ*)		*Megará=**de***	*Megaró=**then***	Megara
TOPO	*Korinthó-**thi***		*Korinthón=**de***	(*ek Kórinth-ou*)	Korinth
TOPO	*Abudó=**thi***		(*es Abud-ón*)	*Abudó=**then***	Abudos
pronoun	(*poũ*)		(*poĩ*)	*pó=**then***	where?

Two remarks have to be made in connection with the data in Table 27. First of all, several of the ternary paradigms are heterogeneous in the sense that not all of the word-forms are construed according to the same principles. Those word-forms which do not contain a clitic appear in brackets. Interestingly, none of the TOPOs has clitics in all three of the cells of the paradigm. Thus, a morphological mismatch applies. In addition, there is a formal differentiation of adverbs and nouns. This differentiation results from the distribution of the clitics. Grey shading marks those cells in which adverbs differ from nouns. Proper adverbs employ the clitic =*thi* to encode the static spatial relation of Place whereas this clitic is not attested with every TOPO-N, TOPO, and pronoun. In lieu of =*thi*, some nouns in Table 27 host the morphological dative (in one case, there is also overabundance since both strate-

gies of representing Place coexist). For the purpose of expressing Goal, the adverbs employ the clitic =se which attaches directly to the stem. In contrast, the nouns take the clitic =de which combines with the morphological accusative of the noun – frequently with morphophonological consequences which may delete the final nasal of the exponent of the accusative in the singular, e.g.,
- *Mégara* 'Megara' → accusative *Megar-án* + clitic =*de* = *Megar-áde* (-n > Ø / __ /d/) 'to Megara')

or result in metathesis of the final sibilant of the accusative marker and the initial dental of the clitic in the plural, e.g.,
- *Athḗnai* 'Athens' → accusative *Athḗn-ās* + clitic =*de* = *Athḗn-ādze* (-sd- > -dz-) 'to Athens'.

The differences of adverbs and nouns are neutralised in the domain of the spatial relation of Source since the clitic =*then* almost always attaches to the stem of its host independent of the word-class membership. Only in the case of the name of the city *Kórinthos* is there no evidence of the use of =*then*. Adverbs and nouns differ from each other insofar as clitics are used for the expression of all three of the spatial relations with adverbs whereas nouns tend to restrict the employment of clitics to the expression of only two of the spatial relations.

As far as we can judge, all TOPOs are potential hosts of the above clitics – most often to express Goal and Source. COMMs, however, do not share this potential generally. Only a small subset of the huge class of COMMs is qualified to take the clitics. Those COMMs which allow for combinations with the clitics are TOPO-NS, i.e., they display conceptual affinities to TOPOs. To determine whether or not the use of the clitics is a common practice of TOPOs, we have checked the Odusseias for occurrences of spatial constructions which involve TOPOs. The construction types we have taken account of are bare TOPOs in the dative, accusative, or genitive, PPs with a TOPO as complement, and TOPOs as hosts of the above clitics. There are exactly fifty TOPO-types in our corpus text which yield an overall turnout of 207 tokens. With 118 tokens (= 57%), the PP-strategy is the predominant choice. Case-inflected bare TOPOs are attested 49 times and thus account for slightly less than a quarter of all cases whereas evidence of the use of clitics stems from forty cases (= 19%). Figure 36 reproduces this uneven distribution schematically.

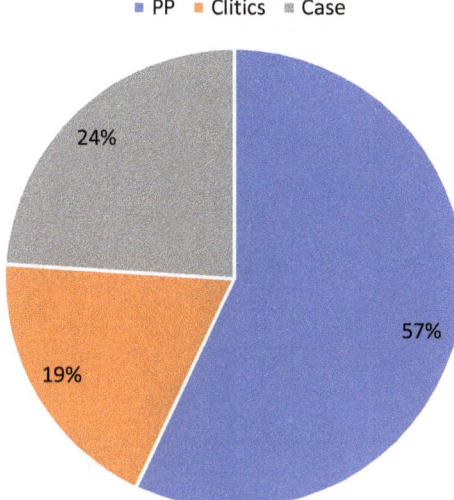

Figure 36: Shares of strategies to encode spatial categories in the Odusseias.

If we take into consideration that the clitic-strategy is banned for good from the expression of the spatial relation of Place, the picture changes considerably to the detriment especially of the strategy of case-inflected bare TOPOS because the exclusion of Place reduces not only the number of tokens to 156 but also that of case-inflected bare TOPOS to 27. At the same time the total of forty tokens of the clitic-strategy remains unaffected and thus its share increases to 26% (with that of the PP-strategy being the same as before, namely 57%, and that of the case-inflected bare TOPOS going down to some 17%). Thus, if we only look at the dynamic spatial relations the significance of the share of the clitic-strategy is high enough to call for closer scrutiny.

There are seven TOPO-types for which the token frequency of the cltic-strategy is higher than that of the competing strategies within the domain of a given spatial category. Most remarkably, the name of the city of Troia is attested seven times in a Goal-construction and ten times in a Source-construction. Five times Goal is expressed by the clitic =*de* and six times Source is expressed by the clitic =*then*. Not nearly as impressive are the turnouts calculated for *Aíguptos* 'Egypt' (Goal: three instances of clitic =*de* vs one instance of the bare TOPO in the accusative), and *Parnāsós* 'Parnass' (Goal: three instances of clitic =*de* without any competing option), and a further five TOPOS, each of which is attested only once in a given spatial relation (viz. Goal *Erebos, Hypereía, Thrāíkē*, Source *Apeira, Ilion*). The TOPO *Púlos* is exceptional in the sense that it provides evidence for each of the competing constructions.[117] Table 28 gives us an inkling of the range of variation. The name of the city occurs 28 times in spatial constructions in the Odusseias – 21 of these attestations involve *Púlos* as Goal. In the bottom row of Table 28 we indicate where in the corpus text the construction can be found (only one example per construction is provided).

Table 28: Shapes of the TOPO *Púlos* in the Odusseias.

Place		Goal			Source	
N_DAT	PP	N_ACC	Clitic	PP	Clitic	PP
1	2	2	7	11	1	4
Pulíoisi	en Púlōi	Púlon	Púlond(e)	es Púlon	Pulóthen	ek Púlou
15, 227	4, 599	3, 4	2, 317	1, 93	16, 323	4, 633

In the Odusseias, there are three TOPOs which give evidence of the clitic strategy for both of the dynamic spatial relations, namely
– *Púlos*: Goal *Púlonde*, Source *Pulóthen*
– *Spártē*: Goal *Spartēnde*, Source *Spártēthen*
– *Troía*: Goal *Troíēnde*, Source *Troíēthen*

Goal is expressed 29 times by way of attaching the clitic =*de* to the TOPO in the accusative. In contrast, there are only eleven instances of Source being expressed by the clitic-strategy. Note, however, that the spatial relation of Source accounts for forty cases of TOPOs in spatial constructions as opposed to 116 cases of Goal, meaning: Source is clearly ousted statistically by Goal independent of the strategy that is used to encode the relations.

It remains to be investigated separately whether or not there are semantic differences which distinguish the competing morphological and morphosyntactic strategies. Since this is an issue that can only be addressed in a dedicated study on an appropriately enlarged philological basis, the best we can do here is to mention in passing that in at least some of the cases the clitic-strategy might be interpreted as encoding a general direction in lieu of a definitive Goal or Source. However, it is very unlikely that this relatively vague spatial interpretation is applicable to all instances of cliticisation. What seems to be a better indicator for the use of the clitic-strategy and its avoidance is the syntactic weight of the NP. Bare TOPOs are eli-

---

[117] There is no example of *Púlos* being used as a bare TOPO in the genitive to encode Source. This construction type is attested only with three TOPOs (*Itháke*, *Troia*, and *Thrinakia*) with a total of four tokens.

gible as hosts of clitics whereas they are not if they are embedded in more complex NPs (cf. above). In the latter case, preference is given to the PP-strategy. Case-inflected nouns alone seem to be admissible under much the same conditions as the clitic-strategy.

In the Odusseias, we have found evidence of COMMs which are subject to the clitic-strategy as, e.g., *āgorá* 'market' → Goal *āgoré̄nde* {Odu 17, 72} (also possible Source *āgoré̆then*), *pólis* 'city' → Goal *pólinde* {Odu 15, 306}. However, the vast majority of these COMMs opt for being the complement of a preposition. Thus, we usually find *eis āgorè̄n* 'to the market' {Odu 8, 12} and *es pólin* 'to the city' {Odu 15, 553} in lieu of the clitic-strategy. Not all of these nouns are also attested with cliticised =*then* for Source. This supposed gap may be caused by the character of our corpus text.[118] However, a random check of the dictionary of Ancient Greek-German (Gemoll 1979) reveals that the gaps cannot be filled easily. We therefore assume that the clitic-strategy was still productive with TOPOs in Homeric Greek when its use on TOPO-NS was already on the decline. The temporary vitality of the above system qualifies TOPOs as a class of their own.

Table 29 summarises the above findings. Boldface is used to highlight those cells from which a given feature is absent. Grey shading marks those cells which host values identical to those of the TOPOs. Single underlining identifies those values which are recessive.

Table 29: Distribution of features over word-classes in Ancient Greek.

	adverbs	TOPOS	TOPO-NS	COMMS
bare DAT / Place	no	yes	yes	(yes)
bare ACC / Goal	no	(yes)	yes	no
clitic / Place	yes	(yes)	no	no
clitic / Goal	=se	=de	=de	no
clitic / Source	=then	=then	(=then)	no

TOPOs and TOPO-NS share the highest number of features although in two cases the equivalence is limited because the features are residual with one of the two classes. Both classes are sandwiched between adverbs and genuine COMMs. Future investigations must determine how large the class of TOPO-NS was in Homer's epoch. Irrespective of the outcome of this future study, it is clear that at an early stage of the history of Ancient Greek TOPOs constituted a class whose members displayed common morphological and morphosyntactic properties which separated them

---

118 To what extent meter might have influenced the presence/absence of the cltics is a problem we cannot address in this study.

from COMMs in general. On this basis, it is possible to characterise the behaviour of Ancient Greek TOPOs to be a relic of a prior period during which the use of prepositions was not yet as widely developed as in later periods of the documented history of the language. Accordingly, TOPOs have retained properties which COMMs had already lost.

Figure 37 tentatively summarises the basic facts of the situation encountred in Ancient Greek. It is understood that we deliberately simplify matters again.

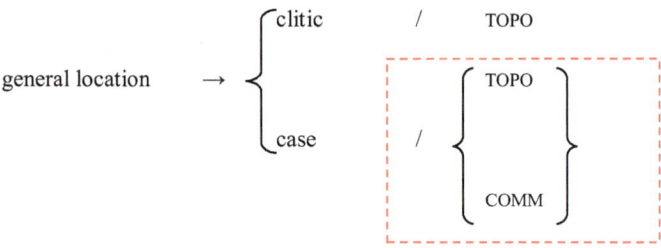

Figure 37: Split rule for general location in Ancient Greek.

In the light of our findings in connection with the non-Indo-European languages discussed in the foregoing sections, it comes as no surprise that in Ancient Greek too the special behaviour of TOPOs is most evident in the domain of the spatial categories. It is to be expected therefore that we encounter a similar situation in Latin whose TOPOs are the topic of the susbequent section.

5.4.2 Latin

An aspect of the grammar of TOPOs in Latin has caught the attention of Anderson (2007: 188) who summarises a particular rule as follows:

> Another kind of morphosyntactic reflexion of the place-name 'hierarchy' is illustrated by the use of the 'terminal accusative' in Latin. Thus, 'the Names of Towns and small islands, when used as limits of Motion Whither, are put in the Accusative' (Gildersleeve and Lodge 1968: 213), while 'Countries and large islands being looked upon as areas, and not points, require prepositions'.

Different kinds of TOPOs receive different kinds of treatment within the grammatical system of Latin as shown in Example (100).

(100) Latin [Indo-European, Latinic] {Ale 1, 1}
*Bello   Alexandrino   conflato   Caesar   **[Rhod-o]***
war:ABL   Alexandrian:ABL   trigger:PTCPL:ABL Caesar   **[Rhodos-ABL]**
atque   [ex   Syri-a   Cilici-a-que]   omnem
and   [from   Syria-ABL   Cilicia-ABL-and]   all:ACC
classem   arcessit.
fleet:ACC   call:PERF.3SG
'After the Alexandrian war had broken out Caesar called the entire fleet **[from Rhodos]** and **[from Syria and Cilicia]**.'

In the very opening sentence of *De Bello Alexandrino liber*,[119] the author employs three TOPOs, one of which refers to an island of moderate size (*Rhodos*) whereas *Syria* and *Cilicia* are names of relatively extended territories in the eastern part of the Mediterranean. All three of the TOPOs are involved in a dynamic spatial relation of Source, i.e., the ships of Caesar's fleet which were stationed in Rhodos, Syria, and Cilicia were ordered to leave for Egypt. Interestingly, the Source relation is expressed by the inflectional ablative alone on the name of the island Rhodos whereas the names of the larger geo-objects Syria and Cilicia are complements of the preposition *ex* 'from, out of' which governs the morphological ablative. Thus, we witness a correlation between the construction type and the ontological class of the reference object in the sense that the spatial relation of Source is represented differently according to the size of the geo-object to which the TOPO refers. If there is rule-governed variation already within the class of TOPOs it does not seem to be too far-fetched to assume that there might also be a formal differentiation of TOPOs and COMMs. This hypothesis can be shown to be correct indeed.

Discounting the vocative, the Latin declension of nouns comprises five case categories, namely nominative, genitive, dative, accusative, and ablative. This set of morphological distinctions is shared by the vast majority of COMMs, ANTHs, and TOPOs. We exemplify this morphological parallelism with the paradigms of three feminine items – the COMM *rosa* 'rose', the ANTH of the goddess *Diana*, and the name of the capital city of the Roman Empire, *Roma* 'Rome' (it is understood that ANTHs

---

[119] In contrast to the practice of restricting our empirical basis to only one classical text of Ancient Greek, we draw examples from a variety of Latin sources, namely, on the one hand, those which are authored by or attributed to Gaius Iulius Caesaer (written between 50 and 44 BC) and, on the other hand, Aurelius Victor's *Liber de Caesaribus* which reflects the Late Latin of the second half of the fourth century of our era. In spite of the huge time gap of about four centuries, the sources yield a remarkably homogenous picture. We thus feel justified in our choice of corpus. It is understood that our Latin case-study needs verifying on a much broader empirical basis.

and TOPOs do not have a plural; as in Ancient Greek, there are also TOPOs which are formally plural like *Athenae* 'Athens'):
- COMM: *rosa* 'rose'
    - singular: nominative *ros-a*, genitive *ros-ae*, dative *ros-ae*, accusative *ros-am*, ablative *ros-ā*[120]
    - plural: nominative *ros-ae*, genitive *ros-arum*, dative *ros-is*, accusative *ros-as*, ablative *ros-is*
- ANTH: *Diana* → nominative *Dian-a*, genitive *Dian-ae*, dative *Dian-ae*, accusative *Dian-am*, ablative *Dian-ā*
- TOPO: *Roma* 'Rome' → nominative *Rom-a*, genitive *Rom-ae*, dative *Rom-ae*, accusative *Rom-am*, ablative *Rom-ā*

The formal identity of the above paradigms glosses over some fundamental differences which come to the fore if we look at the expression of spatial categories.

Luraghi (2009: 298–299) dedicates a section of her study to the morphology of Latin TOPOs. First of all, she emphasises that in contrast to COMMs and ANTHs in general,

> the singular of city names and names of small islands belonging to the first (*-ā*-stems) or second (*-ŏ*-stems) declension, and a few other nouns, retained a separate locative case. Thus, such Latin toponyms were very conservative in that they could occur within spatial expressions without prepositions and continued the tripartite sub-system of Proto-Indo-European. (Luraghi 2009: 298)

With TOPOs of the 1st and 2nd declension, this locative (i.e., the expression of Place) was formally identical to the genitive singular whereas with COMMs Place was usually expressed in the format of a PP headed by the preposition *in* 'in, at' which governed the ablative (Touratier 2013: 185–186), cf. (101)–(103).

(101) Latin [Indo-European, Latinic] – TOPO / Place
    (a) 1st declension / singular {Cae 31, 2}
        Ad quem expugnandum profecti
        to PRO.ACC defeat:GER depart:PTCPL:PL.M
        [*Interamn-ae*] *ab suis caeduntur*
        [Interamna-GEN]$_{PLACE}$ from POSS.3:ABL.PL kill:3PL.PASS
        'Having gone to defeat him they were slain by their (own troops) **[in Interamna]**…'

---

[120] In our sketch of the situation in Latin, we indicate vowel length only in those cases in which our sources provide the information.

(b) 2nd declension / singular {Cae 20, 8–9}
Pescennium Nigrum   apud   Cyzicenos,   Clodium Albinum
Pescennius:ACC Niger:ACC at   Kyzikos   Clodius:ACC Albinus:ACC
[Lugdun-i]   victos   coegit   mori
[Lugdunum-GEN]~PLACE~ defeated:PTCPL:ACC.PL force:PERF.3SG   die:INF
'He forced Pescennius Niger and Clodius Albinus, defeated near Kyzikos and **[in Lugdunum]**, respectively, to die.'

(102) Latin [Indo-European, Latinic] COMM / Place
(a) 1st declension / singular {Cae 20, 6}
Velut Severum   ipsum,   quo   praeclarior
like   Severus:ACC self:ACC as_much   excellent:CMPR
[in republic-a]~PP~ fuit   nemo
[in state-ABL]~PLACE~ be:PERF.3SG nobody
'**[In the country]** nobody was more excellent than Severus himself.'

(b) 2nd declension / singular {Ale 17, 2}
Perfectis   enim   magna   et   parte
finish:PTCPL:ABL.PL   because   big:ABL   and   part:ABL
munitionibus   **[in   oppid-o]**~PP~   et   insulam   et
fortification:ABL.PL   **[in   fortress-ABL]**~PLACE~ and island:ACC and
urbem   uno   tempore   temptari   posse
town:ACC   one:ABL   time:ABL   attack:INF.PASS   be_able.INF
confidebat.
be_confident:IMPERF:3SG
'Since most of the fortifications **[in the fortress]** had been built, he was confident that both the island and the city could be attacked at the same time.'

In (101a), the feminine TOPO *Interamna* comes in a shape that is homophonous to the genitive *Interamn-ae* 'of Interamna' and functions as locative. The same is true of the morphological genitive *Lugdun-i* 'of Lugdunum' of the neuter TOPO *Lugdunum* (i.e., modern Lyon). In contrast to these mono-word constructions, COMMs like *republica* 'state' and *oppidum* 'fortress; village' require the presence of the preposition *in* 'in, at' which governs the ablative when the spatial relation of Place is to be expressed as in (102).

In the case of TOPOs which belonged to the 3rd, 4th, or 5th declension or were formally plural, the locative function was taken over by the morphological ablative. Since the ablative of all declension classes also had the task of expressing Source, there was Place-Source syncretism for a sizable number of TOPOs (whereas others neatly distinguished Place from Source). Goal, on the other hand, was expressed by the prepositionless accusative on TOPOs regardless of the declension class to which they belonged. In (103), the TOPOs *Leptis* and *Tripolis* illustrate the behaviour of the members of the 3rd declension class.

(103) Latin [Indo-European, Latinic] – TOPO / Place = Source ≠ Goal (3rd declension)
    (a) Place {Afr 9, 1}
        [Lept-i] sex cohortium praesidio cum
        [Leptis-ABL]$_{PLACE}$ six cohort:GEN.PL garrison:ABL with
        Saserna relicto
        Saserna:ABL leave_behind:PTCPL:ABL
        'After he had left behind six of the cohorts as garrison under Saserna **[in Leptis]**...'
    (b) Source {Cae 20, 19}
        Quin etiam *[Tripol-i],* cuius
        even also **[Tripolis-ABL]**$_{SOURCE}$ REL:GEN
        [Lept-i] oppido oriebatur,
        [Leptis-ABL]$_{SOURCE}$ town:ABL originate:IMPERF.3SG.PASS
        bellicosae gentes submotae procul.
        belligerent:NOM.PL people:NOM.PL expel:PTCPL:NOM.PL afar
        'Even **[from Tripolis]** – he stemmed **[from Leptis]**, a city thereabouts – the belligerent tribes were driven off.'
    (c) Goal {Afr 62, 5}
        cum primo mane *[Lept-im]*
        with first:ABL morning **[Leptis-ACC]**$_{GOAL}$
        universa classe vectus
        entire:ABL fleet:ABL sail:PTCPL.NOM.SG
        '...early in the morning he sailed with the entire fleet **[to Leptis]**...'

Examples (103a) and (103b) show that the inflectional ablative neutralised the distinction of Place and Source with TOPOs of declension classes other than the 1st/2nd declension (singular). Out of context, word-forms like *Tripoli* 'in/from Tripolis' and *Lepti* 'in/from Leptis' are ambiguous as to their possible Place and Source readings.[121] The accusative (as, e.g., *Leptim* in (103c)), however, does not pose any problems of ambiguity since it is always associated with the spatial relation of Goal. As to COMMs, all three of the spatial relations under scrutiny require PPs independent

---

[121] In (103b), the verb *orior* 'to rise, originate, grow, start' takes a complement NP in the ablative which is not necessarily the Source in the spatial sense of the term. The ablative of the TOPO *Lepti* 'from Leptis' is telling nevertheless because with COMMs as complements, PPs with *ex* 'from; out of' or *a(b)* 'from' are much more common than the mono-word construction.

of the declension class to which the complement noun belongs. In (102b) above, we have shown that the COMM *oppidum* 'fortress; village' requires the presence of the preposition *in* 'in' which governs the ablative on the noun to encode the spatial relation of Place. In (104), the same noun occurs in constructions which also have the format of PPs, namely [*in* $N_{ACC}]_{GOAL}$ and [*ex* $N_{ABL}]_{SOURCE}$. The differential case-government of the preposition *in* 'in; (in)to' is responsible for the existence of syntactic minimal pairs like *in oppido* 'in the town' vs *in oppidum* '(in)to the town'. Similarly, the compatibility of the morphological ablative with different prepositions is at the basis of syntactic minimal pairs such as in *oppido* 'in the town' vs *ex oppido* 'from/out of the town'.

(104) Latin [Indo-European, Latinic] – COMM
 (a) Goal {Afr 6, 3}
 Accidit res incredibilis, ut equites
 happen:3SG.PERF thing incredible that horseman:PL
 minus XXX Galli Maurorum equitum
 less 30 Gaul:PL Moor:GEN.PL horseman:GEN.PL
 II milia loco pellerent fugarentque
 2,000 place:ABL push:3PL.IMPERF.SUBJ flee:3PL.IMPERF.SUBJ:and
 [*in oppid-um*]$_{PP}$.
 [in town-ACC]$_{GOAL}$
 'The incredible thing happened that less than thirty Gaul cavalry drove two thousand horsemen of the Moors from the field and chased them **[into the town]**.'
 (b) Source {Afr 7, 1}
 Legati [*ex oppid-o*]$_{PP}$ *obviam veniunt*
 envoy:NOM.PL **[from town-ABL]**$_{SOURCE}$ towards come:3PL
 'Envoys **[from the city]** come towards him...'

The morphological and morphosyntactic behaviour of TOPOs differs markedly from that of COMMs. Where TOPOs usually express Place, Goal, and Source by the case-inflected noun alone, COMMs are involved in multi-word constructions headed by a spatial preposition.

This observation leads us back to the quote from Anderson (2007) given at the beginning of this section. In this quote, a difference is postulated that divides the class of TOPOs in two. In the above examples, the TOPOs are mostly of the kind which refers to settlements. Names of cities and other settled places are said to form a unitary class together with the names of small islands – a class that is distinct from all other geonyms (and all other kinds of names in general) because of their morphological and morphosyntactic properties sketched above. In point of fact, the names of extended topographic entities such as provinces, countries, large islands, etc. display the same properties as those which we have identified for COMMS, cf. (105).

(105) Latin [Indo-European, Latinic] – TOPOS of larger geo-objects
    (a) Place {Ale 72, 1}
        Zela est oppidum [in Pont-o]$_{PP}$ *positum*
        Zela be.3SG town:NT **[in Pontus-ABL]**$_{PLACE}$ place:PTCPL:NT
        'Zela is a town which is situated **[in Pontus]**...'
    (b) Goal {Gal 4: 20, 1}
        tamen **[in Britanni-am]**$_{PP}$ *proficisci contendit*
        however **[in Britain-ACC]**$_{GOAL}$ depart:INF.PASS strive:3SG.PERF
        '...he however strove to depart **[for Britain]**...'
    (c) Source {Gal 4: 38, 1}
        Caesar postero die T. Labienum legatum
        Caesar subsequent:ABL day:ABL T. Labienus:ACC general:ACC
        cum iis legionibus, quas **[ex**
        with DEM:ABL.PL legion:ABL.PL REL:ACC.PL **[from**
        Britanni-a]$_{PP}$ *reduxerat,* *in Morinos*
        Britain-ABL]$_{SOURCE}$ bring_back:3SG.PLUP in Morinian:ACC.PL
        qui rebellionem fecerant misit.
        REL rebellion:ACC make:3PL.PLUP send:3SG.PERF
        'On the following day, Caesar sent Titus Labienus with those legions which he had brought back **[from Britain]** against the Morinians who had raised a rebellion.'

At the time of the Roman Civil War between Caesar and his opponents, *Pontus* (temporarily unified with neighbouring Bithynia) was a province of the Roman Empire situated on the southern shore of the Black Sea. In the same era, *Britannia* 'Britain' was a large island situated beyond the borders of the Roman Empire. Both geographic units count among the large geo-objects – a classification which manifests itself structurally insofar as for each of the spatial relations, the TOPO of the large geo-object has to be accompanied by an appropriate preposition.

In Table 30, we give a schematic account of the above differences on the morphological and morphosyntactic level. Grey shading highlights the cell of the so-called locative case which is an exclusive trait of only a subset of Latin TOPOS (albeit a sizable one). The names of cities and small islands are privileged morphologically and morphosyntactically since their case-inflection is sufficient to express the spatial relations although syncretism of Place and Source is common with another subset (marked in bold). The dividing line runs right across the class of TOPOS. The names of larger geo-objects correspond to COMMS morphologically and morphosyntactically.

Table 30: TOPO split in Latin.

relation	TOPO			COMM
	city / small island		other	
	1ˢᵗ/2ⁿᵈ declension singular	other		
Goal	ACC	ACC	*in* + ACC	*in* + ACC
Source	ABL	ABL	*ex* + ABL	*ex* + ABL
Place	LOC	ABL	*in* + ABL	*in* + ABL

The idea that there was an additional distinct locative case with some of the TOPOs is based on the fact that no preposition accompanied the TOPO, i.e., the case-inflected word-forms alone served to express the spatial relation of Place. The prepositionless accusative and the prepositionless ablative are different from the locative because the former two remain in place if the appropriate preposition (*in* or *ex*) is added. In contrast, the locative never occurs in combination with any preposition. Were it not for the very small number of words of the 3ʳᵈ and 5ᵗʰ declension which use the exponent *-ī* of the genitive singular of the 2ⁿᵈ declension (Touratier 2013: 187), there would be no reason to treat this case differently from the spatially employed ablative and accusative. The following is almost an exhaustive list of the noteworthy items:

– *domus* 'house' → *dom-(u)ī* **'at** home' [genitive *dom-ūs* / ablative *dom-ō ~ dom-ū* **'from** home'], accusative *dom-um* 'home$_{\text{GOAL}}$'
– *humus* 'ground' → *hum-ī* **'on** the ground' [genitive *hum-ūs* / ablative *hum-ō* **'from** the ground'], (but *in hum-um* 'to the ground'),
– *rūs* 'country' → *rūr-ī* **'in** the country' [genitive *rūr-is* / ablative *rūr-e* **'from** the country'], accusative *rūs* 'to the country', and
– *Carthago* → locative *Carthagin-ī ~ Carthagin-e* [= ablative **'from** Karthago'] **'in** Karthago' [genitive *Carthagin-is*], accusative *Carthagin-em* 'to Karthago'

The syntactic usage of the first of the above cases is illustrated in (106).

(106) Latin [Indo-European, Latinic] – special locative on COMM {Gal 1: 18, 6}
neque solum ***[dom-i]***, sed etiam apud finitimas
and_not only **[house-LOC]**$_{\text{PLACE}}$ but also at neighboring:ACC.PL
civitates largiter posse
community:ACC.PL much be_able.INF
'...(that) he had influence not only **[at home]** but also in the neighbouring communities...'

As to the issue of whether or not the special locative is indeed a supernumerary case characteristic of only a certain subclass of TOPOs and TOPO-Ns, it must be remembered that both ablative and accusative are used on TOPOs without prepositions to indicate Goal and Source (and/or Place), respectively. If the separate locative is formally identical to the genitive singular in the vast majority of cases, one might claim that the prepositionless spatial accusative constitutes a distinct allative. As to the morphological ablative, Luraghi (2009: 297) remarks that with COMMs its dominant function was that of an instrumental whereas on TOPOs it had different spatial functions according to the declension class (as mentioned above, the prepositionless ablative expressed Source with TOPOs of the 1st and 2nd declension but was syncretic as to the distinction of Source and Place with nouns of other declension classes).

Touratier (2013: 187) also mentions the genre-dependent possibility that TOPO-Ns like *locus* 'place' and *regio* 'region' are solely inflected for the ablative without accompanying preposition. This option arises if the TOPO-N is modified by adjectives but not if a determiner accompanies the TOPO-N as shown in Example (107).

(107) Latin [Indo-European, Latinic] – TOPO-N / Place
    (a) with preposition {Gal 5: 7, 3}
        itaque   *dies*   *circiter*   *XXV*   **[in$_{PREP}$ e-o$_{DET}$ loc-o$_N$]$_{PP}$**
        and_so   day:PL   about   25   **[in   DEM-ABL   place-ABL]$_{PLACE}$**
        commoratus
        stay:PTCPL
        '…and so he stayed for about twenty-five days **[in this place]**…'
    (b) without preposition {Gal 5: 49, 7}
        consedit   *et,*   *quam*   **[*aequissim-o*]$_{ADJ}$**
        give_way:3SG.PERF   and   how   **[even:SUPESS-ABL]$_{PLACE}$**
        potest   **[*loc-o*]$_N$**   *castra*   *communit*
        be_able:3SG   **[place-ABL]$_{PLACE}$**   camp   fortify:3SG.PERF
        '…he stopped and **[at the evenest place]** he could find he built an encampment…'

This means that under special conditions some TOPO-Ns assimilate their morphological and morphosyntactic behaviour to that of the TOPOs of the 3rd, 4th, and 5th declension – but only for the expression of the spatial relation of Place. Like in the case of COMMs, Source and Goal require the use of a preposition. In poetry, the inflectional accusative alone can be employed to encode Goal irrespective of the word-class to which the Ground belongs (Touratier 2013: 188). In this study, however, the poetic genre is not taken account of because of its culture-specific traditions and conventions which escape being captured easily in the framework of an approach which focusses on morphological and morphosyntactic patterns of several languages. Moreover, Luraghi (2009: 298) claims that

> already in Early Latin, toponyms of the first two declensions could occur in location expressions with *in* and the ablative

and in a footnote (Luraghi 2009: 298, footnote 23) she adds that

> [i]ndeed the rule by which toponyms did not take prepositions in space expressions was much more consistently followed in the highly artificial language of Classical writers than in early Latin.

Accordingly, our above account of the situation in Latin reflects exclusively the properties of what is characterised as an "artificial" version of the language. However, we consider Classical Latin and its Late Latin successor to be fully-blown varieties of Latin, the structural traits of which constitute a legitimate object of linguistic analysis since the patterns we have identified are rule-governed no matter how distant from spoken Latin the written register happened to be. More importantly, Luraghi (2009: 298) argues that the TOPOs preserved a diachronic stage that COMMs had already left behind in the course of the 2nd century BC. The distinctive features of the TOPOs are conservative traits.

ANTHs form a class of their own in the sense that firstly, they do not allow for the use of prepositionless constructions and secondly, they employ only a subset of those prepositions which are compatible with (inanimate) COMMs to express the dynamic spatial relations, namely *ad* 'to' (in lieu of *in* 'in') and *a(b)* 'from' (in lieu of *ex* 'from, out of'), cf. (108).

(108) Latin [Indo-European, Latinic] – ANTH
    (a) Goal {Ale 26, 3}
        Inde   re   bene   gesta   [Alexandre-am]
        then thing:ABL well   carry:PTCPL:ABL **[Alexandria-ACC]**$_{GOAL}$
        [ad   Caesar-em]$_{PP}$ *contendit*
        [to   Caesar-ACC]$_{GOAL}$ hurry:3SG.PERF
        'After having concluded the affair successfully he marched hurriedly **[to Caesar] [in Alexandria]**.'
    (b) Source {Gal 4: 12, 1}
        quod   legati   eorum   paulo   ante
        since envoy:PL DEM:GEN.PL little:ABL before
        [a   Caesar-e]$_{PP}$ *discesserant*
        [from Caesar-ABL]$_{SOURCE}$ separate:3PL.PLQ
        '...because their envoys had departed **[from Caesar]** only shortly before...'

Example (108a) is especially telling because it provides evidence of the prepositionless accusative with the TOPO which occurs side by side with the PP that is necessary to encode the same spatial relation of Goal with an ANTH.

ANTHs resemble COMMs insofar as members of both classes employ PPs to express spatial relations. However, the sets of prepositions that are made use of are different. COMMs are split into two groups. Inanimate COMMs have two options for each of the dynamic spatial relations, namely, in the case of Goal, either *in* or *ad* with the accusative and, in the case of Source, *ex* or *a(b)* with the ablative. These two pairs of prepositions reflect the distinction of interior and exterior spatial relations (i.e., illative vs allative and elative vs ablative). Animate COMMs[122] and ANTHs behave as one. Both preferably take the prepositions of the exterior relations.[123] This means that on the supposed continuum of nouny word-classes, Latin ANTHs are not located between TOPOs and COMMs but wind up at the far end, i.e., at a distance from the TOPOs, cf. Table 31.[124]

Table 31: Continuum of shared properties in Latin.

criterion	TOPOS			TOPO-N	COMM / inanimate	COMMS / animate; ANTH
	small island / city		other			
	1st/2nd declension / sg	other				
Prep_PLACE	no	no	*in*	(*in*)	*in*	*apud*
Prep_GOAL	no	no	*in*	*in*	*in*	*ad*
Prep_SOURCE	no	no	*ex*	*ex*	*ex*	*a(b)*
locative	yes	no	no	no	no	no

---

122 Cf. examples like: (a) Place – {Gal 1: 50, 4} . . .*quod [apud Germanos]_PP ea consuetudo esset*. . . '. . .that it is a costum **[among the Germanic tribes]**. . .', (b) Goal {Gal 5: 58, 1} . . .*ut nulla ratione ea res enuntiari aut [ad Treveros]_PP perferri posset*. '. . .so that this event was betrayed or could reach **[the Treveri]**.' (note that the construction [in + ethnonym/animate N_accusative] usually conveys the meaning of 'against', cf. [47.3] above), (c) Source – {Gal 7: 71, 7}. . .*pecus, cuius magna erat copia [a Mandubiis]_PP compulsa*. . . '. . .cattle which they had sequestered in great numbers **[from the Mandubii]**. . .' (but cf. again {Gal 6: 9, 1} . . .*postquam [ex Menapiis]_PP [in Treveros]_PP venit*. . . '. . .after he had come **[from the Menapii] [to the Treveri]**. . .'). In this study, we cannot account for the relatively frequent instances of variation. Our generalisations therefore have to be taken with a grain of salt.
123 But cf. e.g. Source – {Gal 4: 23, 5} . . .*quae [ex Voluseno]_PP cognovisset*. . . '. . .what he had learned **[from Volusenus]**. . .'
124 We have not found any uncontroversial evidence of a Place relation of the inessive brand with an animate Ground in the texts of our Latin corpus. What can be found in great numbers, however, are cases like {Gal 1: 33, 2} . . .*eorumque obsides esse [apud Ariovistum]_PP*. . . '. . .and (that) their hostages were **[at Ariovist's]**. . .' with the preposition *apud* 'at' (which governs the accusative) which invokes an adessive reading.

Grey shading identifies those values which are singularities in a given row. Boldface is used to mark those values which are identical to that of TOPOS of small islands and cities of the category OTHER. The use of identical prepositions is indicated by underlining. Since *in* 'in' is optional with TOPO-NS under certain conditions, the preposition is put in brackets without underlining. This peculiarity would also justify a change of place. The column of TOPO-NS could be put between the two categories OTHER under the heading TOPOS. This modification is tantamount to strengthening the split that affects the TOPO-class because TOPO-NS would then be more similar to the names of small islands and cities than other TOPOS are.

Figure 38 cannot do justice to all of the phenomena reflected in Table 31 and discussed in the foregoing paragraphs. Simplification is once more of the essence to make our point.

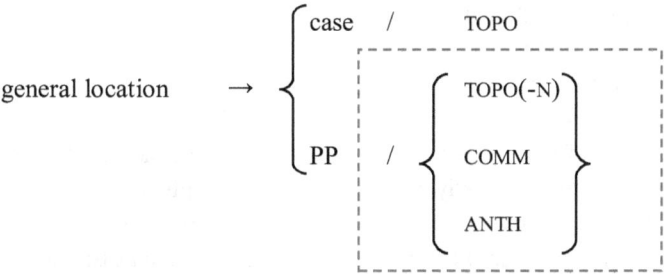

Figure 38: Split rule for general location in Latin.

In Latin, TOPOS do not form a homogeneous class. A segment of this class is singled out because of its special behaviour on the morphological and the morphosyntactic level. This heterogeneity goes hand in hand with morphological mismatches which can be characterised as instances of periphrasis and/or antiperiphrasis as well as syncretism (i.e., when the morphological ablative is used for both Place and Source). The formal identity of the expressions of Place and Source is especially remarkable because, in the literature on the grammar of space, this constellation is considered to be highly marked if not downright impossible (as discussed in Lestrade 2010b: 96–105). The fact that TOPOS provide fertile ground for the survival or the genesis of markedness renders their study especially interesting for general linguistics.

5.4.3 Conservatism and structural complexity

The data from Ancient Greek and Latin tell us various things. First of all, we learn that the special status of TOPOS within the inflectional system of the two classic languages is tightly connected to what one might term conservatism. The TOPOS have retained morphological properties which have disappeared from most other paradigms. The latter are representative of morphological and morphosyntactic innovations whereas TOPOS clearly lag behind in the overall process of language change. This also means that the distinctive grammar of TOPOS is secondary since it is an epiphenomenal effect of say, diachronic slow motion of this particular class of nouny elements. In the absence of large-scale quantitative data, we are not in a position to verify whether or not the diachronic slowness of Ancient Greek and Latin TOPOS can be connected causally to their type and/or token frequency.

Second, the preservation of older stages of the inflectional system with TOPOS in both of the languages studied in Section 1.4 means that the morphosyntactically simpler mono-word constructions of the TOPOS contrast with the innovative multi-word syntagms or PPs that have to be used with COMMS and ANTHS to express the above spatial relations. One might interpret this contrast of morphosyntactically relatively simple vs morphosyntactically relatively complex constructions to reflect some kind of markedness hierarchy. In this specific case, it could be argued that, in an untechnical manner of speaking, the association of spatial relations with TOPOS is more natural than that of spatial relations with ANTHS and COMMS so that the higher naturalness licenses the less costly expression on the morphosyntactic level.

This still somewhat speculative assumption receives support from the third observation that can be made in connection to the data from Ancient Greek and Latin, namely that the morphological peculiarities which justify the separation of certain sub-classes of TOPOS from other nouny classes come to the fore exactly in the functional domain of the spatial categories. This is in line with what we have seen in Lezgian, Hungarian, Finnish, and Basque before: TOPOS and the grammar of space (Svourou 1993) are interconnected in a particularly intimate way. The raison d'être of TOPOS as a distinct word-class is rooted in this special relation between TOPOS and the grammar of space.

5.5 Swahili

We complement the empirical part of this section with a discussion of the relative data from the Bantu language Swahili. In contrast to all other previous languages (as discussed in the foregoing subsections of Section 5), Swahili is not equipped with morphological case. However, there is bound morphology that is outwardly reminiscent of inflectional case. In their chapter on spatial adverbials of Swahili, Möhlig and Heine (1999: 64) explain that

> [d]ie deutsche Präposition 'in' wird im Swahili entweder durch **katika** [. . .] oder durch die Partikel **-ni** wiedergegeben. Diese wird als Nachsilbe dem Nomen, das als Ortsbezeichnung verwendet werden soll, angefügt. [. . .] Sofern es sich nicht um geographische Eigennamen [Ortsnamen, Ländernamen etc.] handelt, muß jede Ortsbezeichnung, die auf **iko** 'ist, befindet sich' bzw. auf **haiko** 'befindet sich nicht' folgt, um die Nachsilbe **-ni** erweitert werden.[125] [original boldface]

The authors circumscribe a situation of rule-governed variation. There are several ways to express the static spatial relation of Place in Swahili. The Ground-noun can either be the complement of the preposition *katika* 'in'[126] or it can host the synonymous suffix *-ni* unless it is a TOPO. In addition, in combination with the locative copula *iko* 'to be somewhere' or its negated variant *haiko* 'not to be somewhere' all Ground-nouns are required to host *-ni* – except TOPOs. On this basis, we can assume that the use of the preposition *katika* 'in' is barred for clauses, the nucleus of which is one of the above copulas. If, however, *katika* 'in' is excluded from this context and TOPOs usually do not combine with *-ni*, then it must be the case that TOPOs are zero-marked for the spatial relation of Place as adverbials of *iko* / *haiko*. This means that we have to reckon with at least three structural options to express Place as shown in Figure 39.

$$\text{Place} \rightarrow \begin{cases} katika \ / \ X_{\text{(ha)iko}} & \begin{cases} \text{COMM} \\ \\ \text{TOPO} \end{cases} \\ \text{-}ni & / \ ((ha)iko) \ \text{COMM} \\ \emptyset & / \ (ha)iko \ \text{TOPO} \end{cases}$$

Figure 39: Three options of encoding Place in Swahili (preliminary version).

---

[125] Our translation: "The German preposition 'in' is rendered in Swahili either by **katika** [. . .] or by the particle **-ni**. This is added as a suffix to the noun that is to be used as a place name. [. . .] Apart from geographical proper names [settlement names, country names, etc.], every place noun that follows **iko** 'to be somewhere' or follows **haiko** 'not to be somewhere' must be extended by the suffix **-ni**."
[126] This preposition has been grammaticalised from the (relational) noun *kati* 'centre', the old genitival form of which is reflected by *katika* 'in' (Brauner and Herms 1986: 288, footnote 87).

As will become clear shortly, this catalogue of options needs to be refined since a further possibility is not mentioned as such in our reference works of Swahili but can be identified on the basis of corpus data (cf. below).[127]

Before we address the existence of a fourth construction type, it is in order at this point to provide empirical evidence of the structural options which are covered by the rule in Figure 39. In (109), two synonymous constructions are presented which involve an inanimate COMM in the function of Ground. In combination with the locative copula *-mo* 'to be in sth.', which forms a paradigm of three with *-po* 'to be at/on/near something' and *-ko* 'to be at a distance from something' (Brauner and Herms 1986: 207), the inanimate COMM may either be embedded in a PP or the host of *-ni*.[128]

(109)   Swahili [Atlantic-Congo, Bantu] (Brauner and Herms 1986: 299)

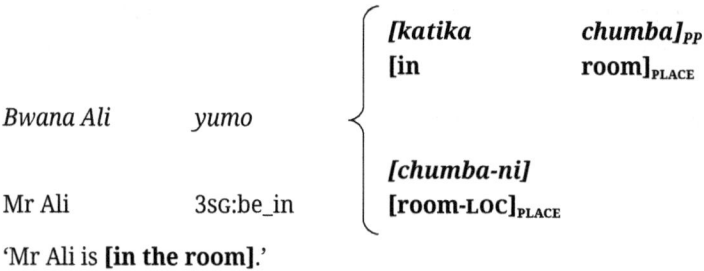

'Mr Ali is **[in the room]**.'

---

127 We do not go into the issue of the so-called locative classes which have survived as agreement categories in Swahili. The prefixes *mu-* 'in', *pa-* 'at, on, near to', and *ku-* 'at a distance from' are no longer used on (head) nouns but occur obligatorily on attributes, anaphors, determiners, relative pronouns, and verbs provided the head noun hosts the locative marker *-ni* as in *[Soko-ni]*₍PLACE₎ *ni ku-zuri* 'It is nice [on the market].' Brauner and Herms (1986: 197) emphasise that TOPOS which generally do not take the locative marker *-ni* trigger the same agreement phenomena as in *[Tanzania]*₍PLACE₎ *ni ku-zuri* 'It is nice [in Tanzania].' Since TOPOS do not behave differently from COMMS in this domain, we skip discussing the issue at length. The interested reader is referred to Harries (1977).
128 The varieties of Swahili which are at the basis of, on the one hand, the didactic material by Shivutse (1972), Brauner and Herms (1986), and Möhlig and Heine (1999) and, on the other hand, the translation of *Le Petit Prince* and the Swahili electronic corpus HCS are not absolutely identical structurally in the segment of grammar that concerns us in this study. For obvious reasons, we refrain from fathoming the depth of the normative grammar of a transnationally employed language which is not the native language of the vast majority of its speakers in several countries in East Africa. Therefore, we take at face value the examples our sources provide no matter how much the sources (dis)agree among each other.

For TOPOs, the second possibility, i.e., the suffixation of the spatial marker *-ni*, is ruled out generally. Nevertheless, TOPOs too allow for two different constructions for the purpose of expressing Place. The choice of construction depends on the kind of predicate nucleus of the clause in which the TOPO is involved. If this function is fulfilled by one of the locative copulas, no relator is used at all as in (110).

(110) Swahili [Atlantic-Congo, Bantu] (Brauner and Herms 1986: 299)
Bwana Ali yuko   [Tanzania].
Mr Ali   3SG:be_at_distance [Tanzania]$_{PLACE}$
'Mr Ali is (down there) **[in Tanzania]**.'

In contrast, zero-marking of Place is possible not only for TOPOs but also for inanimate COMMs in combination with the static verb *ishi* 'to live somewhere'. However, there is variation since the PP-strategy is also attested with TOPOs as Grounds as in (111).

(111) Swahili [Atlantic-Congo, Bantu]
  (a) zero-marked TOPO (Brauner and Herms 1986: 83)
    Bwana Müller   anaishi   [Leipzig].
    Mr Müller   3SG:PRS:live [Leipzig]$_{PLACE}$
    'Mr Müller lives **[in Leipzig]**.'
  (b) PP-strategy with TOPO (Brauner and Herms 1986: 69)
    Ninaishi   pamoja   na   mama   [katika Leipzig]$_{PP}$.
    1SG:PRS:live together with mother [in   Leipzig]$_{PLACE}$
    'I live together with my mother **[in Leipzig]**.'

Example (112) from the Swahili translation of *Le Petit Prince* shows that the same verb combines with an inanimate COMM without intervening preposition but with the spatial marker *-ni* affixed to the Ground-noun.

(112) Swahili [Atlantic-Congo, Bantu] – affixation of *-ni* / COMM {LPP Swahili, 64}
    Kama watu   bilioni mbili   wanaoishi   [dunia-ni]
    when CL1.PL:human billion CL1:two 3PL:PRS:REL:live [world-LOC]$_{PLACE}$
    'If the two billion people who live **[on the Earth]**...'

We have not found a single instance of an inanimate COMM which functions as Ground-noun in combination with *ishi* 'to live somewhere' that does not overtly mark that it is involved in a spatial situation. This failure is perhaps attributable to the small size of our Swahili corpus.

If the verb in a clause is neither a locative copula nor the static verb *ishi* 'to live somewhere', TOPOs which have the role of Ground in a given context need to be accompanied by the preposition *katika* 'in' as in (113).

(113)   Swahili [Atlantic-Congo, Bantu] (Brauner and Herms 1986: 63)
    (a) PP-strategy with TOPO
        [Katika Tanzania]$_{PP}$ watu   wote   wana
        [in   Tanzania]$_{PLACE}$ CL1.PL:human   CL1.PL:all   CL1.PL:with
        watoto.
        CL1.PL:child
        '**[In Tanzania]** everybody has children.'
    (b) PP-strategy with inanimate COMM [different predicate nucleus from (51)]
        Pamoja   *naye*   *walimu*   *kumi*   *wanafundisha*   *[katika*
        together   with:3SG   CL1.PL:teacher   ten   3PL:PRS:teach   [in
        shule]$_{PP}$.
        school]$_{PLACE}$
        'Together with him/her ten teachers are teaching **[in the school]**.'

There is thus an area of overlap in which TOPOs and inanimate COMMs behave similarly. The distribution profile of the two word-classes overlaps outside the context of the locative copulas because both TOPOs and inanimate COMMs opt for the PP-strategy. At the same time, in combination with the locative copulas, TOPOs and inanimate COMMs behave markedly differently from each other in terms of the constructions they occur in. This is what one can observe in relation to the expression of Place. In the next paragraph, we check whether or not there are also structurally relevant differences in connection to the expression of the spatial relations of Goal and Source.

Brauner and Herms (1986: 61) take up this issue when they state that

[n]ach Verben der Bewegung wie *-fika* 'ankommen', *kwenda* 'gehen', *-toka* 'herkommen' u.a. [...] folgt die Ortsangabe unmittelbar dem Verb und nicht als Umstandsbestimmung des Ortes mit Präposition wie im Deutschen.[129]

In accordance with the explanations given in the German quote, many motion verbs in Swahili are immediately followed by the Ground-noun without intervening preposition. This is shown for TOPOs as Ground-nouns in (114).

---

[129] Our translation: "After motion verbs such as *-fika* 'to arrive', *kwenda* 'to go', and *-toka* 'to come from', etc. [...], the indication of place follows the verb directly and not as an adverbial of place with a preposition as in German."

(114) Swahili [Atlantic-Congo, Bantu] – dynamic spatial relations / TOPOS
    (a) Goal / TOPO (Möhlig and Heine 1999: 73)
        anawasili   **[Dar-es-Salaam]**   leo   *alasiri*
        3SG:PRS:arrive   **[Dar-es-Salaam]**$_{GOAL}$ today   afternoon
        '…he will arrive this afternoon **[in Dar-es-Salaam]**…'
    (b) Source / TOPO (Möhlig and Heine 1999: 67)
        Ninatoka   **[Ujerumani]**.
        1SG:PRS:come_from   **[CL6:Germany]**$_{SOURCE}$
        'I come **[from Germany]**.'

This means that TOPOS display the same behaviour in constructions which express dynamic spatial relations as in the above combinations with the locative copulas in Place-constructions, i.e., zero-marking applies. In the case of inanimate COMMS as Ground-nouns in Goal-constructions and Source-constructions, the situation is different. In contrast to TOPOS, zero-marking is not an option for COMMS. Some kind of overt marker is compulsory for inanimate COMMS, namely either -*ni* as in (115a) or the preposition *katika* 'in' as in (115b).

(115) Swahili [Atlantic-Congo, Bantu] – dynamic spatial relations / COMMS
    (a) Goal / COMM (Möhlig and Heine 1991: 91)
        Hamisi   *anataka*   *kwenda*   **[shule-ni]**.
        Hamisi   3SG:PRS:want   go   **[school-LOC]**$_{GOAL}$
        'Hamisi wants to go **[to the school]**.'
    (b) Source / COMM (Brauner and Herms 1986: 289)
        Anatoka   **[*katika*   *chumba*]**$_{PP}$.
        3SG:PRS:come_from   **[in   room]**$_{SOURCE}$
        'He is coming **[out of the room]**.'

Both of the above sample sentences require a comment. As to (115a), there is evidence also of zero-marking in combinations of the same verb and inanimate COMM as Ground-noun as, e.g., in (116).

(116) Swahili [Atlantic-Congo, Bantu] – Goal / COMM (Brauner and Herms 1986: 291)
    Alikwenda   **[shule]**.
    3SG:PRET:go   **[school]**$_{GOAL}$
    'He went **[to (the) school]**.'

Brauner and Herms (1986: 290–292) present this example in the context of their discussion of the multi-purpose preposition *kwa* (which, for simplicity, we gloss as equivalent of English *at*), the spatial functions of which will occupy our minds further below.¹³⁰ For the time being it suffices to note that (116) does not contain any preposition, be it *kwa* or other. What is more, the Ground-noun *shule* 'school' is completely bare, i.e., the expected spatial marker *-ni* is absent too. Is the presence of *-ni* in (115a) and its absence in (116) an instance of free variation? We hypothesise that it is not, although, in the absence of further examples and without the appropriate contextual information, our assumption rests on shaky ground. We suggest that of the two examples under scrutiny, only (115) is properly spatial in the sense that it describes a situation in which a human Figure executes a motion in space from A to B in order to reach a concrete Goal (identified by the Ground-noun *shule*). This genuinely spatial nature of the situation requires an overt marker – in this case the appropriate *-ni* to yield *shuleni*. Example (116) on the other hand describes a situation of a different kind. What is asserted in (116) is the fact that the referent of the subject-prefix attended school or received schooling in the past. This example does not describe a concrete motion event at all and that is why we put the definite article in brackets in the English translation of (116).

Postulating a formal distinction along the lines of the opposition concrete vs abstract fits Brauner and Herms' (1986: 289) observation that

> [b]ei einer allgemeinen Feststellung, wo vom konkreten räumlichen Bezug abstrahiert wird, entfällt *katika*. [. . .] Es ist zu beachten, daß auch *kutoka* ohne *katika* verwendet wird, wenn allgemein die Herkunft gemeint ist.¹³¹

The more concrete the spatial situation the higher the probability that an overt spatial marker will be used. Or the other way round: abstractness correlates with the absence of an overt spatial marker (with inanimate COMMs as Ground-noun). Superficially, Example (115b) runs counter to the hypothesis of Brauner and Herms (1986: 61) according to which the presence of the ablative motion verb *-toka* 'to come from' precludes the use of the preposition *katika* 'in'. If we understand Brauner and Herms (1986: 289) correctly this rule applies only if the situation is not genuinely

---

130 In point of fact, *kwa* is grammaticalised from the genitive of an erstwhile locative class with the prefix *ku-* (Brauner and Herms 1986: 202).
131 Our translation: "In a general statement where concrete spatial reference is abstracted, *katika* is omitted [. . .]. It should be noted that *kutoka* is also used without *katika* if origin is the intended meaning."

spatial.[132] If the spatial character of the situation is genuine, *katika* 'in' may even cooccur with other so-called prepositions such as *kutoka* 'from',[133] cf. (117).

(117) Swahili [Atlantic-Congo, Bantu] – Source / COMM (Brauner and Herms 1986: 289)

Maaskari	*walikuja*	**[*kutoka***	**[*katika***	*kijiji*	*cha*
CL4.PL:soldier	3PL:PRET:come	**[from**	**[in**	CL3:city	CL3:of

jirani]_{PP}]_{PP}
CL3:neighbor]_{PLACE}]_{SOURCE}
'The soldiers came **[from the neighbouring village]**.'

According to Brauner and Herms (1986: 289), *kutoka* 'from' can be used without the additional *katika* 'in' if the general origin is expressed, i.e., if reference is not being made to a concrete motion event in the framework of a genuine spatial situation. This differential behaviour of *kutoka* and *katika* can be captured by assuming a general abstract-concrete opposition.

The preposition *kutoka* forms pairs with other prepositions in constructions which indicate the point of departure and the endpoint of a given motion event. According to Brauner and Herms (1986: 290), there are three such paired constructions, namely
- [*kutoka* NP *hadi* NP] – termed "allgemein" (general), i.e., any kind of NP can be used in this construction which is not restricted to spatial situations in the first place (cf. ***kutoka** hapa **hadi** jengo hili* '**from** here **to** this building'),
- [*kutoka* TOPO *kwenda* TOPO] – this construction tolerates exclusively TOPOs as fillers of the NP-slots (cf. ***kutoka** Bagamoyo **kwenda** Kizimkazi* '**from** Bagamoyo **to** Kizimkazi'[134]),

---

132 Brauner and Herms (1986: 61 and 289) illustrate the absence of relators in the case of abstract, i.e., not properly spatial situations with (i) *Anatoka **[Moshi]***. 'S/he comes (= stems) **[from Moshi]**_{SOURCE}.' and (ii) *Anatoka **[Tanzania]***. 'S/he comes (= stems) **[from Tanzania]**_{SOURCE}.' However, these examples are not entirely convincing since they involve TOPOs as Ground-nouns for which zero-marking in combination with motion verbs is assumed to be general anyway. At the same time, Examples (i)-(ii) do not describe concrete spatial situations but inform about the place of origin or birth of an individual. They provide primarily biographical information in lieu of giving an account of a motion event.
133 The preposition *kutoka* 'from' is homophonous with and has been grammaticalised from the infinitive of *-toka* 'to come from', i.e., from *kutoka* 'to come from'.
134 The preposition *kwenda* 'to' is formally identical and has been grammaticalised from the verb *kwenda* 'to go'.

– [*kutoka* TOPO *kuelekea* TOPO] – as far as can we can judge from the information provided in our reference text, this construction is synonymous to the previous option and thus should only allow for TOPOs, cf. Example (118).[135]

(118) Swahili [Atlantic-Congo, Bantu] – Source and Goal (Brauner and Herms 1986: 290)
Ndege inaruka **[*kutoka* Entebbe]**$_{PP}$ **[*kuelekea* Kairo]**$_{PP.}$
bird CL2:PRS:fly **[from Entebbe]**$_{SOURCE}$ **[to Kairo]**$_{GOAL}$
'The plane is flying **[from Entebbe] [to Kairo]**.'

There is thus a further context in which TOPOs receive a morphosyntactic treatment that differs from that of inanimate COMMs. This special treatment manifests itself in the existence of two constructions which admit only TOPOs as fillers of the NP-slots as opposed to a third construction which can be said to neutralise the COMM-TOPO distinction since it is open to members of all nouny word classes discussed so far.

In sum, TOPOs are unmarked for Goal whereas an inanimate COMM always cooccurs with an overt marker which identifies them as belonging to a spatial construction.[136] This overt marker, however, does not specify which spatial relation applies since the markers are sensitive neither to the distinction of static vs dynamic nor to the dimension of directionality, i.e., the notions of Place, Goal, and Source are neutralised. Their disambiguation is largely achieved via the semantics of the predicate nucleus. Furthermore, TOPOs boast of special constructions of the Source-Goal kind from which inanimate COMMs are excluded. All these phenomena speak clearly in favour of the separate status of TOPOs in the sense that they cannot be subsumed sweepingly under the same heading as inanimate COMMs.

The question remains whether what we have observed in the above paragraphs is distinctive of TOPOs or holds for all kinds of names such as ANTHs. To

---

135 *Kuelekea* 'to' has the shape of an infinitive as the prefix *ku*- suggests. Thus, it can be assumed that we witness another case of grammaticalisation of a preposition from a verbal source. However, we have not been able to identify the appropriate verb itself since neither of our sources lists *-elekea* or anything remotely similar in their vocabularies.

136 We assume that the variation of the PP-strategy and the use of *-ni* is common also with the dynamic spatial relations. Brauner and Herms (1986: 61), for instance, state that the motion verb *ingia* 'to enter' always requires the presence of the preposition *katika* 'in' as in (i) *Mwalimu anaingia* **[*katika darasa*]**$_{PP.}$ 'The teacher enters **[into the classroom]**$_{GOAL}$.' However, there is evidence of an alternative in Möhlig and Heine (1999: 121), namely (ii) . . . *Juma na Ali waliingia* **[*msitu-ni*].** '. . . Juma and Ali went **[into the forest]**$_{GOAL}$.' For the purpose at hand, it is only of secondary importance to determine whether this variation is general or the reflexion of dialectal differences. Irrespective of which strategy is employed, the fact that the dynamic spatial relation of Goal is overtly marked with inanimate COMMs remains unaffected by the variation.

approach this subject-matter it is necessary to note that in Swahili, COMMs are subject to an animacy-based distinction. That is why we have insisted hitherto on characterizing the class of COMMs as that of the inanimate brand. From Brauner and Herms (1986: 291) we learn, for instance, that in Goal-constructions, the appropriate preposition is the multifunctional *kwa* 'at' (cf. above) if the Ground-noun is [+human] as in (119).

(119) Swahili [Atlantic-Congo, Bantu] – Goal / human COMM (Brauner and Herms 1986: 291)
Alikwenda **[*kwa profesa*]**$_{PP}$.
3SG:PRET:go **[at professor]**$_{GOAL}$
'He went **[to the professor]**.'

When we compare (119) to (116), we immediately notice that there is a contrast as to the overt vs covert expression of the spatial relation. We already know that the absence of the expected preposition *katika* or of the spatial marker *-ni* in (116) might be causally connected to the more abstract character of the utterance which does not refer to a concrete spatial situation. Nevertheless, the presence of the preposition *kwa* in (119) is indicative of general differences between inanimate and animate (or more precisely, human) COMMs in Swahili. If the spatial relation of Goal is expressed overtly, inanimate COMMs and human COMMs opt for different markers. Where inanimate COMMs have the choice of *katika* and *-ni*, human COMMs always take *kwa*.

Our pedagogical grammars of Swahili do not explicitly say so but the examples in (120) indicate that human COMMs are different from non-human COMMs not only as to their behaviour in connection to the spatial relation of Goal but also more generally in all instances of general location, be it static or dynamic.[137]

(120) Swahili [Atlantic-Congo, Bantu] – further spatial categories with human COMMs
(a) Place / human COMM (Shivutse 1972: 68)
Twakaa **[*kwa kaka yetu*]**$_{PP}$
1PL:HABIT:live **[at older_brother CL1:1PL.POSS]**$_{PLACE}$
'We live **[at our older brother's place]**.'

---

[137] The fact that there are several areas in the grammatical system of Swahili where the split between human and non-human reference has repercussions on the structural level is mentioned explicitly repeatedly for instance by Brauner and Herms (1986: 131, footnote 43, 149, etc.) though never in connection with the topic at hand.

(b) Source / human COMM (Shivutse 1972: 128)
Msichana huyu anatoka **[*kwa shangazi***
CL1:girl DEM:CL1 3SG:PRS:come [at aunt
yake]_{PP}
CL1:3SG.POSS]_{SOURCE}
'This girl is coming **[from her aunt('s place)]**.'

(c) Source / human COMM (Brauner and Herms 1986: 287)
lililofungwa na Adam Sabu baada ya kupata
CL4:PRET:REL:shoot:PASS with Adam Sabu after INF:receive
pasi nzuri **[*kutoka* [*kwa mchezaji machachari***
pass CL5:good [from [at CL1:player CL1:restless
Abdallah Kinadeni]_{PP}]_{PP}
Abdallah Kinadeni]_{PLACE}]_{SOURCE}
'(. . .a goal which was) scored by Adam Sabu after he had received a good pass **[from the indefatigable player Abdallah Kinademi]**.'

The possibility of *kwa* being used in constructions which express the spatial relation of Place – as in (120a) – is evident also from the following fact. According to Möhlig and Heine (1999: 228), the pronominal forms of *kwa* are used in a way that is reminiscent of the French preposition *chez* 'at someone's place' since they can be translated into German as *bei X zu Hause* 'at X's place/home'. This means that they are admissible in contexts which can be characterised as static, i.e., in spatial situations which correspond to that of Place. The paradigm contains the following word-forms:

– *kwa* + PRONOUN → *kwangu* 'at my place', *kwako* 'at your$_{SG}$ place', *kwake* 'at his/her place', *kwetu* 'at our place', *kwenu* 'at your$_{PL}$ place', *kwao* 'at their place'.

Examples (120b)–(120c) give evidence of *kwa* being used in constructions which express the dynamic spatial relation of Source. In both of the examples, a form of *toka* 'to come' is involved. In (120b), it comes in the shape of the finite motion verb *anatoka* 's/he is coming' whereas in (120c) the preposition *kutoka* 'from' is the head of a PP which embeds a second PP headed by *kwa*. As in the previous cases, the outer preposition *kutoka* is responsible for encoding directionality. The inner preposition *kwa*, however, is neutral as to directionality and thus expresses location and indirectly the high degree of animacy of the Ground-noun.

Given that this animacy-based differentiation is grammatically pervasive, it suggests itself to assume that ANTHs which usually have human referents also behave differently from inanimate COMMs. Does this mean that they also diverge from TOPOs in terms of morphology and morphosyntax? Do they side formally with human COMMs or do they constitute a class of their own? The examples in (121) prove that ANTHs and human COMMs are similar to each other as to their morphosyntactic behaviour.[138]

(121) Swahili [Atlantic-Congo, Bantu] / ANTH
    (a) Place (Shivutse 1972: 128)
           Rafiki   yangu   kutoka   Tanganyika akaa   [kwa
           friend   CL1:1SG.POSS INF:come Tanganyika 3SG:live [at
           *Bwana Müller]*_{PP}
           Mr Müller]_{PLACE}
           'My friend from Tanganyika lives **[at Mr Müller('s place)]**.'
    (b) Goal {Dun 9}
           Kasim   alipotoka   nyumbani   alikwenda
           Kasim   3SG:PRET:REL:come CL9:house:LOC 3SG:PRET:go
           *moja kwa moja **[mpaka   [kwa   Mzee Rehani]_{PP}]_{PP}**
           directly   **[until   [at   Mzee Rehani]_{PLACE}]_{GOAL}**
           'When Kasim exited the house, he went directly **[to Mzee Rehani]**.'
    (c) Source (Doc: 626915 Corpus: hcs_kiongozi)
           *Nimekumbuka   wewe   unatoka   **[kwa   Makamba]**_{PP}*
           1SG:PERF:remember you   2SG:PRS:come **[at   Makamba]**_{SOURCE}
           'I remember that you come **[from Makamba]**.'

In each of the three sentences, the Ground-noun is an ANTH which is obligatorily accompanied by the all-purpose preposition *kwa* no matter what directionality applies. In (121b), *kwa* is the inner preposition which combines with the outer preposition *mpaka* 'until'. The general similarities of the behaviour of the human COMMs in (119)–(120) are thus unquestionable. The shared high degree of animacy defines ANTHs and human COMMs as a separate class from non-human COMMs on the one hand and TOPOs on the other hand.

On account of the above discussion, we are in a position now to refine the rule given in Figure 39 above. To refine the preliminary picture, it is necessary to distinguish human from non-human COMMs. In addition, ANTHs and human COMMs can be united to form the class of human nouns. This means, of course, that there is no general class of names since TOPOs and ANTHs go separate ways morphosyn-

---

[138] Since ANTHs are scarcely attested in the functions as Ground-nouns in our conventionally published sources, the evidence for ANTHs in constructions which express the dynamic spatial relations of Goal and Source are taken from the Swahili electronic corpus. We are grateful to our colleagues Susanne Hackmack and Karl-Heinz Wagner for searching the corpus for these and further examples to the benefit of this study.

tactically, cf. Figure 40. In contrast to its predecessor, Figure 40 is meant to cover all spatial relations in lieu of only that of Place as in Figure 39.

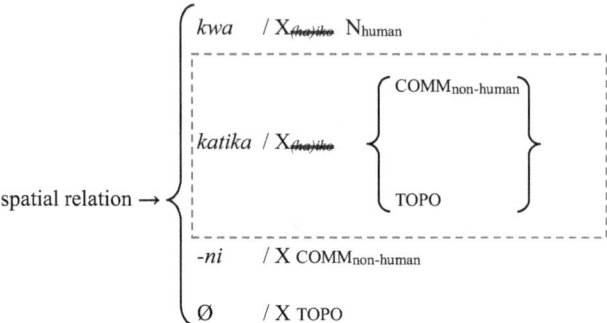

Figure 40: Four options of encoding spatial relations in Swahili.

Each of the classes of nouny word categories associates with an encoding strategy of its own. Except for the occasional evidence of TOPO-NS in combination with a closed set of verbs, zero-marking of spatial relations is the privilege of TOPOS. Non-human COMMS and human nouns express spatial relations overtly. With human nouns, the appropriate preposition is *kwa* which is insensitive to directionality. Similarly, directionality is also underspecified in the case of the locative marker *-ni* which is reserved for non-human COMMS. If we discount the preposition *katika* which is admissible in combination with non-human COMMS as well as with TOPOS, the above strategies are distributed in such a way that one is reminded of classifiers, in this case more precisely of locative classifiers (Aikhenvald 2000: 172–183). Most interestingly, Swahili TOPOS are exempt from cooccurring with an overt locative classifier since they are already inherently spatial and thus license zero-marking. As in the previous case-studies, TOPOS are most dissimilar from other nouny word-classes exactly in the domain of the spatial categories for which they seem to be particularly well suited. In terms of morphology and morphosyntax this means that TOPOS require the structurally simplest constructions to express the spatial relations, which, in turn, makes it plausible that zero-marking can be employed especially (if not exclusively) with TOPOS.

5.6 Preliminary results: *Special Toponymic Grammar* in spatial relations

In Sections 5.1–5.5, we have investigated a number of STG phenomena in the domain of spatial relations in selected languages. Glossing over a small number of details, it is possible to generalise as follows:

(a) If there are differences in the structural complexity of constructions which express spatial relations, those constructions which involve TOPOS tend to be less complex than those of COMMS and/or ANTHS. This applies to Basque, Ancient Greek, and Latin where the constructions involving a TOPO count at least one morphological component less than their counterparts which involve COMMS or ANTHS.

(b) More specifically, if one or several of the spatial relations can or must be zero-marked, zero-marking usually associates with TOPOS. This is the case in Swahili where only TOPO(-N)S are licit as bare locative adjuncts in contrast to COMMS and ANTHS which always require the presence of a dedicated marker for spatial relations.

(c) On the other hand, if there is evidence of morphological singularities, i.e., highly specialised bound markers with restricted distribution, these singularities more often than not are attested exclusively with TOPOS. This is what we see in the retention of the erstwhile locative cases in Latin and Hungarian as well as in the employment of the clitics in Ancient Greek.

(d) Moreover, to express spatial relations, subsets of TOPOS are entitled to a choice of markers which differs from that of COMMS and/or ANTHS, i.e., the rules which hold for the bulk of the nouny word-classes are invalid in the domain of TOPOS. We witness this scenario in Lezgian and the Uralic languages.

The aspects of STG presented in (a)–(d) do not form a unitary pattern because the behaviour of TOPOS deviates from that of COMMS and/or ANTHS in multiple and diverse ways. Nevertheless, (a)–(d) are in line with our expectations formulated earlier in this study. By and large, our working hypothesis (27b) is borne out by the empirical facts. We confidently assume that the situations sketched in the previous subsections will be observable also in many other languages which have not been inquired into in this pilot-study. As the project TYPTOP continues, further evidence of parallel morphological and morphosyntactical behaviour of TOPOS will be collected to prove that the above case-studies are not exceptional but representative of a great many languages world-wide.

We repeat for emphasis that the special behaviour of TOPOS is strongest in the domain of the grammar of space. This fact hardly surprises us because TOPOS are at home in the realm of spatial relations, in a manner of speaking. Simplifying, it can be assumed that TOPOs usually fulfil the function of spatial adverbials, i.e., they occur as Ground-nouns in constructions which convey information about the Place, Goal, or Source of (participants who are involved in) an event. In contrast, COMMs and ANTHs do not figure as prominently as TOPOs in this domain. ANTHs are much better candidates for the status of the internal or external argument of verbs whereas COMMs display similar preferences, especially in relation to the syntactic function of (direct) object (to use a traditional term for once). We elaborate on these issues in Section 6. For the time being it suffices to state that all these observations boil down to assuming the unmarked status of TOPOs in the grammar of space. At the same time, it can be assumed that outside the grammar of space, TOPOs constitute an alien element – a status that can also be expected to be reflected by their behaviour.

The data from Basque, the Northeast Caucasian language Lezgian, the three Uralic languages Estonian, Finnish, and Hungarian, the Indo-European languages Ancient Greek and Latin as well as from the Bantu language Swahili strongly support the idea that we are dealing with a cross-linguistically recurrent pattern which privileges TOPOs structurally in the functional domain of the expression of spatial relations. The pattern shows up in languages which are not only genetically unrelated (with four different major language families and an isolate) but also typologically (ergative vs accusative languages, SOV vs SVO, prefixing vs suffixing, agglutinating vs fusional) and areally sufficiently diverse (different parts of Europe and Africa). One might argue that so far we have looked exclusively at languages which are equipped with rich inflectional morphology (not necessarily implying the existence of morphological case) and thus the evidence we have adduced is perhaps typical only of languages which boast elaborate systems of bound morphology. However, the interaction of adpositional phrases with mono-word constructions as discussed for Lezgian, Ancient Greek, Latin, and Swahili is already suggestive of the possibility that similar phenomena which single out TOPOs as a distinct word-class are also observable beyond inflectional morphology.

Given that these characteristic traits of STG are cross-linguistically recurrent, it is striking to see that hitherto they have raised sufficient interest neither within the framework of the Grammar of Names nor in that of the Grammar of Space.[139] Especially with regard to the latter, this neglect is surprising because the research programme dedicated to spatial categories in language and cognition has grown into a major interdisciplinary endeavour in which many scholars from many parts of the

---

[139] We acknowledge that neither of the two research programmes constitutes a unitary and homogeneous approach. Both the Grammar of Names and the Grammar of Space are labels for internally diversified spheres of academic interest, in a manner of speaking.

world are actively involved (Levinson 2003). To promote the cause of STG by way of making proponents of the Grammar of Space aware of the scientific potential with which TOPOS are invested we highlight a selection of special points of interest in the remainder of this section – special points of interest which suggest that TOPOS and STG should be given more prominence in studies on spatial language in the future. Further aspects will be presented in Section 6.

A good starting point for the subsequent discussion is Langacker (2008: 73) who argues that "[e]ven at the conceptual level, the objects of our mental universe have no inherent status as profile, trajectory, or landmark." For the sake of the argument, we pick out the last of the three categories, viz. landmark. Landmark is terminologically akin to the notion of Ground whereas trajectory and Figure can be equated with each other (Goldap 1991: 13–14). Figure 41 captures the basics of a spatial situation employing the terminology introduced to the discussion by Talmy (1985).

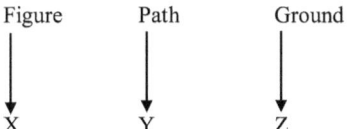

Figure 41: Main components of spatial situations.

In the construal of a spatial situation, the participants differ from each other on the parameter of prominence. Landmarks are considered to be associated with secondary focus (Langacker 2008: 70–71). Since a landmark "is the entity with respect to which the Trajector is located" (Svorou 1993: 9) the linguistic expressions which serve to verbalise the landmark have the function of Ground. The terminological (semi-)equivalence is one thing. More important, however, is the de natura candidature of TOPOS for the function of Ground. As we have seen throughout the foregoing subsections TOPOS are overwhelmingly involved in constructions in which they fill the slot reserved for the Ground. Not only do they preferably function as Ground but when fulfilling this function TOPOS frequently display grammatical properties which are not regularly shared by either ANTHs or COMMS.

Nyström (2016: 44) defines the onomasticon as the "proprial part of the mental lexicon." The existence of a toponomasticon in the sense of a geo-referential mental lexicon for TOPOS is not discussed but can be inferred from the context. The toponomasticon functions as a reservoir not only of the TOPOS as such but also of everything that is associated with them including their eligibility for Ground-hood. What renders TOPOS additionally interesting for the Grammar of Space is their pronounced tendency to provoke mismatches by way of opting for structural patterns other than those which are normally chosen by ANTHs and COMMS.

A particularly striking case is illustrated by example (110) from Swahili. In a still undetermined but in any event sizable number of languages, TOPOs may come as naked syntactic words when functioning as spatial adjuncts, i.e., zero-marking applies. This cross-linguistically recurrent fact can be taken to mean that TOPOs are Grounds by default because they need not be formally established for this function. Only if they are used outside this functional domain do they require special morphological means to signal that they are used for different syntactic purposes. We go a step further and challenge Langacker by way of claiming that TOPOs most probably do have the inherent status of landmark on the conceptual level. In support of this strong claim, we refer to Stolz and Levkovych (2020a) who present cases from a variety of languages spoken in the Americas and Irian Jaya (Western New Guinea) in which TOPOs themselves are morphologically complex insofar as they obligatorily host a locative affix. This is why the referential or naming function of the TOPO is formally identical with the TOPO used as Ground in spatial situations (especially when it comes to expressing Place). The tenability of the admittedly daring assumption of the conceptually anchored Ground-hood of TOPOs calls for being looked into more closely in the future.

The case-studies in Section 5 have shown and were also meant to prove that zero-marking does not exhaust the STG phenomenology. The grammatical expressions needed to encode the function of TOPOs in a given construction are often less complex (but still phonologically realised) in comparison to those necessary for members of other nouny word-classes. Basque is a valid piece of evidence in favour of our above hypothesis although the spatial relations are marked overtly also on TOPOs. What is crucial is the absence of a class marker which identifies TOPOs as what they are. This is zero-marking of a different kind. In the other cases of coding asymmetries, the connection to the supposed inherent Ground-hood of TOPOs is not as obvious. One might want to argue that the lesser degree of complexity of the constructions which involve TOPOs as Ground as opposed to the more complex constructions with ANTHs or COMMs as Ground iconically reflects the basic affinity of TOPOs and the Ground function which lacks in the case of the other nouny word-classes. There is an alternative way to account for the facts which will be discussed in Section 6.

The Uralic languages and Lezgian attest to a different phenomenon. A subset of TOPOs selects a series of spatial cases which is different from that used for COMMs (and ANTHs). This differential behaviour is restricted to qualified TOPOs, namely (many of) those which refer to geo-objects (here: settled places) which historically belong to the territory the speech-community assumes to be rightfully theirs independent of contemporary borders. For convenience, we have simplifyingly indexed TOPOs of this kind with [heritage] as opposed to [foreign] for the majority of TOPOs whose choice of cases is the same as that of COMMs (and ANTHs). Furthermore, the differential behaviour exclusively applies if properly spatial relations have to be expressed. At least superficially, inherent Ground-hood cannot be invoked as an explanation for this behaviour. Explanatory models which are based on coding efficiency are facing a riddle, too. In Section 6, we will try to find a solution to this problem.

TOPOs are not completely absent from research dedicated to spatial language. However, they are mostly mentioned only in passing and almost always only in connection to zero-marking as the references given in Stolz et al. (2014: 17–18). There are indirect references to TOPOs, for instance, in Svorou (1993: 16) when she discusses the technical notion of region, i.e., the focused part of a given landmark. Countries, continents, cities, and sundry geo-object classes are mentioned to determine whether they have regions or are regions. Their representation by TOPOs goes unmentioned. What is more, the geo-objects are listed alongside other entities such as fields, buildings, rooms, phone booths, etc. When lumped together with unnamed entities, the specifics of TOPOs automatically submerge under a plethora of features, properties, and traits associated with other concepts so that the identification of STG becomes impossible.

We conclude this section by way of referring back to Figure 41. The conceptual level connects to the linguistic level where three slots call for being filled. The fillers are represented by the variables X, Y, and Z. The underlying assumption is that no matter which concrete elements occupy their appropriate slots the tectonics of the situation remains unaffected. The immunity of the conceptual grid against whatever happens on the lower levels is most probably responsible for the lack of interest in TOPOs as fillers of slot Z. However, while the elementary configuration of Figure, Path, and Ground is never challenged, it is still a fact that it makes a considerable difference in many languages whether Z is filled by a COMM, an ANTH, or a TOPO. If the choice of Z-filler determines the morphosyntactic shape of the construction which expresses the spatial situation, then the necessity of thoroughly studying the differential behaviour of the different Z-filler classes is unavoidable. If TOPOs in Ground-function trigger or allow for the use of constructions which are structurally less complex than or formally different from those employed for Grounds of a different kind, a conclusion that imposes itself on the analyst is that we are facing either the prototype or a very special kind of spatial configuration. On account of this (perhaps only vague) possibility alone, TOPOs and STG deserve to be scrutinised in as detailed a way as possible also from the vantage-point of the Grammar of Space.

6 Functional-typological evaluation

In Section 4.5 and 5.6, we have presented some preliminary results based on our case-studies on SAG and STG, respectively. These preliminary results need to be integrated into the wider picture of SOG adopting the functional-typological perspective. To this end, we will not only refer back to the examples given Sections 1–5 but also present fresh data from languages which hitherto have not been taken account of in this study. The evaluation in functional-typological terms will be carried out in direct confrontation with additional empirical evidence for SOG. For a start, Section 6.1 compares the grammaticalisation of ANTH-markers in two unrelated languages. The parallels identified for Catalan and Kaqchikel guide us to a better understanding of the impact socio-pragmatic factors have on SOG. Coding efficiency and concepts related to this notion are addressed in Section 6.2 with a view to determining in how far it is possible to explain SOG in terms of structural economy. Section 6.3 is dedicated to the ability of ANTHs and TOPOs to express grammatical categories differently from the patterns which apply in the case of COMMs. The final evaluative section bears the number 6.4 and looks at SOG from the perspective of Canonical Typology.

6.1 A socio-pragmatic long distance parallel: Catalan and Kaqchikel

For language typology, the recurrence of a given phenomenon in languages of different genetic affiliation and different geographical location is of importance. If similar structures are attested in otherwise unrelated languages, possible explanations for the parallel are shared structural preconditions and/or identical functional motivation. TOPOs have many parallels of this kind on offer so that STG in general invites typologists to explore the grammar of PROPs. The cross-linguistic parallels become even more interesting when they can be shown to result from (perhaps still unfinished) processes of grammaticalisation. One such parallel links Catalan [Indo-European, Romance] spoken in Spain and Kaqchikel [Mayan] spoken in Guatemala. Caro Reina (2014: 181) traces the diachronic development of the so-called onymic articles of Catalan whose employment is mandatory in the Balearic varieties of the languages whereas the mainland dialects seem to allow for variation including absence of the marker (Caro Reina 2014: 182–183). The binary system consists of *en* for ANTHs referring to male persons and *na* for female ANTHs. The examples in (122) show that these ANTH-markers are distinct from the gender-sensitive definite articles of Catalan.

(122) Catalan [Indo-European, Romance] (Caro Reina 2014: 179)
   (a) male ANTH
       En   *Ferrer*   *té*   *molta*   *feina*
       ANTH.M   Ferrer   have.3SG   much:F   work:F
       'Ferrer has a lot of work.'
   (b) masculine COMM
       El   *ferrer*   *té*   *molta*   *feina*
       DEF.M   smith   have.3SG   much:F   work:F
       'The smith has a lot of work.' [O.T.]
   (c) female ANTH[140]
       Na   *Rosa*   *és*   *molt*   *maca*
       ANTH.F   Rose   be.3SG   very   beautiful:F
       'Rose is very beautiful.' [O.T.]
   (d) feminine COMM
       La   *rosa*   *és*   *molt*   *maca*
       DEF.F   Rose   be.3SG   very   beautiful:F
       'The rose is very beautiful.' [O.T.]

The above examples are especially convincing since the ANTHs *Ferrer* and *Rosa* are homophonous with the COMMs *ferrer* 'smith' and *rosa* 'rose' from which they historically derive. If the segmental chains are used as ANTHs, they trigger the use of the markers *en* and *na* whereas their COMM-equivalents require the presence of the definite article *el* and *la*, respectively. (122a) and (122b) as well as (122c) and (122d) form minimal pairs. *En* and *na* are thus not to be confounded with the usual definite articles co-occurring with PROPs. As a matter of fact, the definite articles *el* and *la* derive historically from Latin *ille* DEM.PRX.M and *illa* DEM.PRX.F 'this', respectively. As to *en* and *na*, the etymological origin can be traced back to the Latin forms of address *domine* and *domina* (vocatives of the COMMs *dominus* 'lord' and *domina* 'lady') which underwent phonological reduction during the Vulgar Latin and Old Catalan periods (Caro Reina 2014: 190–194). The grammaticalisation channel links the source concept TERM OF ADDRESS to the target concept ANTH-marker. In the second edition of the *World Lexicon of Grammaticalization* (Kuteva et al. 2019: 278 and 459–460), the nearest equivalent to the Catalan case one can find is the grammaticalisation of generic expressions for man and woman to classifiers for male and female (human) beings. This, however, is not what the Catalan example is about because the classifiers registered in the *World Lexicon of Grammaticalization* regu-

---

140 In the original, example (122c) is erroneously identical with (122d) although the English translations differ. We have replaced the definite article *la* with the ANTH-marker *na* in (122c).

larly combine with (animate) COMMs whereas the Catalan markers are reserved for combinations with ANTHs. Since Catalan is cross-linguistically not unique in grammaticalising terms of address to ANTH-markers, a future third edition of the *World Lexicon of Grammaticalization* should feature an additional entry covering cases of this kind.

This future entry should also mention the Mayan language Kaqchikel because it displays striking similarities with Catalan. The normative grammar (Academia de Lenguas Mayas de Guatemala 2006: 104) assumes a morphologically invariable definite article *ri* which is used with COMMs as in *ri tz'i'* '**the** dog' (cf. Becker 2021: 140–145). In contrast, ANTHs are usually preceded by a marker from a set of seven, namely *xta, a, ya, te, ma, nan,* and *tat* which are termed "clasificadores personales" (i.e., personal classifiers) by the Academia de Lenguas Mayas de Guatemala (2006: 141). The markers *tat* and *nan* are used in direct address and independently as pronoun-like titles or COMMs combinable with the definite article *ri*. Similarly, *xat* and *a* can be used in direct address to adults of the same age as the speaker. When used in direct address the markers are always sandwiched between the free pronoun of the 2nd person singular *rat* 'you' and the ANTH (Academia de Lenguas Mayas de Guatemala 2006: 142–143). The markers distinguish the sex and age of the named person as can be gathered from (123).

(123) Kaqchikel [Mayan]
    (a) female ANTH – young (Academia de Lenguas Mayas de Guatemala 2006: 141).
Ye'-ok   *apo*   *la*   *äk'*   *r-ik'in*   *la*   ***xta***
INCMP.ABS.3PL-enter   DIR   DEM   hen   ERG.3SG-RN   DEM   **ANTH.F**
Tz'ununya'
Tz'ununya'
'The hens enter (at the place) where Tz'ununya' is.'
    (b) male ANTH – young (Academia de Lenguas Mayas de Guatemala 2006: 141)
x-oj-b'e   *r-ik'in*   ***a***   *Lu'*
CMPL-ABS.1PL-go   ERG.3SG-RN   **ANTH.M**   Lu'
'We went with Lu'.'
    (c) female ANTH – adult (Academia de Lenguas Mayas de Guatemala 2006: 142)
atux   *chi*   *samaj*   *n-Ø-i-b'än*   *r-ik'in*
INT   PREP   work   INCMP-ABS.3SG-ERG.2PL-do   ERG.3SG-RN
ri   ***ya***   *Ixmukane'*   *k'a?*
DEF   **ANTH.F**   Ixmukane'   PTCL
'What kind of work are you doing with (**Doña**) Ixmukane'?'

(d) male ANTH – adult (Academia de Lenguas Mayas de Guatemala 2006: 142)

X-Ø-pe	ri	**tat**	Lu'
CMPL-ABS.3SG-come	DEF	**ANTH.M**	Lu'

'(**Don**) Lu' came.'

Rodríguez Guaján (1994: 151) assumes the system of ANTH-markers for Kaqchikel which we reproduce in Table 32.

Table 32: System of ANTH-markers in Kaqchikel.

	male	female
young	*a*	*ixta*
married	*tat*	*nan*
elderly	*ma*	*te*

As in the case of Catalan, there is variation. First of all, the marker *ya* for elderly women as a synonym for *te* is not recognised by the author. Second, in some parts of the Kaqchikel speech-territory *ma* is used for bachelors only (Rodríguez Guaján 1994: 152) whereas Brown et al. (2006: 204, 230) assume that *ma* is developing into an honorific ("term of respect"). A third aspect of interest is the tendency for *ixta* to be omitted because speakers of local Spanish have borrowed *xta* into their language where it acquired negative connotations (Rodríguez Guaján 1994: 152). The normative grammar quoted above seems to fix rules to stabilise the system.

The parallel with Catalan comes to the fore when we look at the etymology of the Kaqchikel ANTH-markers. Rodríguez Guaján (1994: 152) identifies the following source-target pairs:

– *tata'aj* 'father'   > *tat*
– *te'ej* 'mother'   > *te*
– *mama'aj* 'grandfather' > *ma*
– *ixtän* 'young woman'   > *ixta*
– *ala'* 'youngster, child' > *a*

The author fails to identify a source for the ANTH-marker *nan* for married women but it suggests itself to connect it to Spanish *nana* 'lullaby, grandmother' which coexists also as vocative *nána* 'ma'am, older woman' in Kaqchikel (Brown et al. 2006: 234). It is uncertain whether the ANTH-marker *ya* (synonymous with *te*) for elderly women originates from Spanish *doña*. As in the Catalan case, the ANTH-markers derive from terms of address. In the process of grammaticalisation, their segmental chains underwent reduction. Given that the initial /i/ of *ixta* was lost between the mid-1990s and the first decade of the new millennium to yield *xta*, all erstwhile polysyllabic COMMs have become monosyllabic ANTH-markers. Except for *nan* and *tat*, the short forms can no longer be used as COMMs. Phonetic attrition and decategorialisation count among the usual concomitants of grammaticalisation (Lehmann 1995: 126–132). In contrast to Catalan, the Kaqchikel ANTH-markers are sensitive to the parameters of age and marital status, i.e., they are semantically more informative than their Catalan counterparts. The dimensions of politeness and affection also come into play because

> [l]los dos últimos clasificadores personales **a** y **xta** masculine y femenino respectivamente, además de diferenciar edad y sexo son clasificadores de afecto y confianza [original boldface] (Academia de Lenguas Mayas de Guatemala 2006: 191).[141]

The degree of grammaticalisation is perhaps higher in Catalan than it is in Kaqchikel. On older stages of Kaqchikel (documented in dictionaries compiled by Spanish missionaries in the early 17th century), most of the lexical items from which the contemporary ANTH-markers developed were attested as COMMs such as *al* 'child (of a woman)', *mama* 'old man', *tata* 'father', *te* 'aunt' (Smailus 1989b: 33, Smailus 1989c: 533, 752–753). In the case of *ixtän* 'young woman', it suggests itself that this word contains the prefix *ix-* which indicates female sex (Smailus 1989a: 106–107). The stem *tän* could not be identified in the lexicographic material provided by Smailus (1989b–c). *Nan* and *ya* are absent from the earliest sources. For those items which are attested already 500 years ago, no sign of grammaticalisation to ANTH-markers is discernible. The grammaticalisation is thus probably a relatively recent development dating to the colonial period or might even be posterior to it. For over half a millennium, Kaqchikel has been exposed to ever-increasing Spanish influence. The Mayan language shares this age-long exposure to pressure from Spanish with Catalan.

Finding similarities in the SAG of unrelated languages is one thing. Another thing is to make sense of these similarities. We emphasise that the parallel cannot be explained with reference to structural homologies between the two languages. Kaqchikel and Catalan disagree on several major typological parameters. Kaqchikel is a VOS language whereas Catalan displays pragmatically variable SVO word-order. Kaqchikel has an ergative or split intransitive alignment as opposed to the accusative alignment of Catalan. Catalan is equipped with grammatical gender whereas Kaqchikel only marks sex with human participants without agreement phenomena. That both languages are heavily inflecting – Kaqchikel leaning towards agglu-

---

[141] Our translation: "the latter two personal classififiers *a* and *xta* masculine and feminine, respectively, besides distinguishing age and sex are classifiers of affection and confidence."

tination whereas Catalan has a stronger fusional component – is of no avail. The same negative judgment holds for the fact that both languages are pro-drop and prepositional languages. The morphosyntax of the ANTH-markers of the two languages is not the same. From (122), we know that in Catalan, ANTH-marker and definite article are incompatible with each other because they occupy the same syntactic slot. In Kaqchikel, however, the ANTH-marker is also licit inside the DP as shown in (124).

(124)   Kaqchikel [Mayan] (Academia de Lenguas Mayas de Guatemala 2006: 243)
        x-Ø-tzij-o-n   chi   r-ij   ri   cholq'ij
        CMPL-ABS.3SG-word-VR-APASS   PREP   ERG.3SG-about   DEF   calendar
        [ri   ma   Sinakan]$_{DP}$
        [DEF   **ANTH.M** Sinakan]
        '[(**Don**) Sinakan] spoke about the calendar.'

The definite article *ri* and the different ANTH-markers are thus no competitors for the same syntactic slot. The co-occurrence of definite articles and ANTH-markers is neither obligatory nor infrequent. Since Spanish too allows for the co-presence of definite articles and titles as in (125) (Bosque 2009: 1258–1259), we do not rule out the possibility that contact with Spanish has reinforced the grammaticalisation process in Kaqchikel in the sense that the frequent use of terms of address of the appositive title kind in Spanish has been copied by bilingual Kaqchikel-Spanish speakers.

(125)   Spanish [Indo-European, Romance] (Bosque 2009: 1259)
        Ha   llamado   [el   **señor**   Francisco García]$_{DP}$
        AUX.3SG   call:PTCPL   [DEF.M   **mister**   Francisco García]
        'S/he has called [**Mr.** Francisco García].'

The constant contact with Spanish notwithstanding, combinations of definite article and ANTH-marker do not seem to be possible in Catalan.

Another interesting aspect associated with Kaqchikel is the absence of a proper system of classifiers. Polian (2017: 219) claims that

> [m]ost Mayan languages are classifier languages and have sets of classifiers that are used, obligatorily or optionally, in quantifying constructions, especially with numerals, a situation that has been reconstructed for Proto-Mayan.

Smailus (1989a: 129–130) reports that in 17th century Kaqchikel cardinal numerals in attributive construction obligatorily hosted a classifier suffix. In the mid-18th century, Flores (1753: 333–344) counted 29 different classifiers. In contemporary

Kaqchikel, however, this system has disintegrated completely owing primarily to Spanish influence according to Rodríguez Guaján (1994: 110). Given that our above assumption about the partly contact-induced nature of the grammaticalisation of ANTH-markers is correct, Spanish would be responsible at least in part for the disappearance of the inherited system of numeral classifiers and, at the same time, also for the rise of a system of ANTH-markers called personal classifiers in the literature on Kaqchikel grammar.

We assume that the Kaqchikel-Catalan parallel has nothing to do with structural correspondences between the Mayan and the Romance language. The motivation of the grammaticalisation of erstwhile forms of address to ANTH-markers is external to grammar in the strict sense. Socio-pragmatic factors, first and foremost those which are related to politeness (Allerton 1996), are responsible for the obligatorification of the use of terms of address in combination with ANTHS. Their co-occurrence rate was high enough to give rise to collocations which in turn were subject to semantic bleaching and phonetic reduction of the term of address which lost its referential functions to become a grammatical marker which now accompanies ANTHs on a regular basis. The grammaticalisation processes both in Catalan and Kaqchikel are rooted in communicative patterns of social interaction in the societies to which the respective speech-communities belong. The socially-grounded overuse of the terms of address in combination with ANTHs was meant to avoid face-threatening acts (Brown and Levinson 1987). Taavitsainen and Jucker (2016: 428–430) discuss a politeness continuum which inter alia involves honorific titles and other titles and their relation to the concept of face. As we see it, one still has determined where exactly on this continuum grammaticalised ANTH-markers can be placed. The rise of the ANTH-markers is functionally motivated since the markers fulfil a socio-pragmatically important function. In terms of economy, the grammaticalisation of the ANTH-markers in Catalan and Kaqchikel is unnecessary and inefficient since they require the permanent use of additional morphological material. In contrast, their phonological erosion and desemanticisation fit the usual correlation of high frequency in use and loss of substance on the expression side.

PROPS are culturally valuable concepts which are crucial for interactions in society (Ainiala 2016) and identity issues (Aldrin 2016). Culture-related factors of this and similar kinds are difficult to handle exclusively in the context of language typology whereas they are largely familiar to onomasticians. There is thus a good reason for the different disciplines to co-operate, not least because we assume that future research will unearth further cases of the kind discussed in this section. We do not claim that every ANTH-marker originates from a former term of address but chances are that cross-linguistically this grammaticalisation channel is activated more often than it might seem. ANTH-markers are no genuine articles. The dissociation of the two categories from each other is reflected by the different grammaticalisation channels they pass through. We assume that TOPO-markers are grammaticalised via a third distinct grammaticalisation channel whose source concepts will be identified in a follow-up study.[142]

6.2 Predictable and expected

In this section, we turn our attention to the explanatory power of typically functionalist notions such as predictability and expectedness to see whether they adequately capture the driving forces behind the structural differences of ANTHs and/ or TOPOs as opposed to COMMs. Coding asymmetries are in the foreground of the subsequent discussion which starts with the most extreme form of asymmetrical coding, namely the contrast between zero-marking and material marking of grammatical categories.

In *The Story of Zero*, Givón (2017: 3) gives two examples of the communicative logic of linguistic zero expressions, namely that predictable and unimportant information "need not be mentioned." Givón's study is dedicated to anaphora and related phenomena. However, the communicative principles are intended to hold generally so that they also hold for the morphosyntactic aspects of SOG. In the foregoing sections, zero-marking has been mentioned time and again in connection with SAG and STG. In the latter domain, TOPOs which function as locative adverbials have been shown to be exempt from overtly marking the spatial relation which connects them to the predicate provided general location applies. Croft (2022: 187) argues for instance, that

> [i]f an argument lacks an overt flag (i.e., the flagging strategy is zero coding), then it almost always expresses a core role. The primary exceptions to this generalization are found when the oblique argument's semantic role is predictable – for example, when a place name is used to describe a location.

The initial sentence of this quote will occupy us further below. At this point, however, we focus on the second part of the quote since TOPOs are explicitly mentioned. Croft (2022: 187) refers to studies by Comrie (1986: 86–88) and Rodrigues Aristar (1997) where similar arguments are put forward. In these studies, it is shown that Eastern Armenian [Indo-European, Armenic] has four different strategies to encode Place with zero-marking being reserved for those cases "when the filler is a place and

---

142 Nübling (2005) cursorily discusses a range of PROP-markers and their diachrony also from languages other than German but the concept of *Eigennamenmarker* (PROP-marker) embraces many phenomena outside the domain of morphosyntax so that not all of the examples fall under the rubric of PROP-marker applied in this study.

the state of affairs has a salient locative role, such as 'live'" (Croft 2001: 235). Note that zero-marking is also possible for Goal but in this case, there is no restriction to TOPOs functioning as Ground (Stolz et al. 2014: 76–80). Lestrade (2010b: 140–146) presents examples of zero-marking of spatial relations from several unrelated languages, many of which come with a COMM as Ground. He states that

> [t]he general observation is, that sometimes, spatial case does not have to be used if the speaker thinks the meaning contribution it would make is sufficiently clear already. (Lestrade 2010: 140)

The gist of his line of argumentation is that "[s]patial case is not used when the spatial meaning is predictable" (Lestrade 2010: 143) and

> "[t]he speaker prefers the most economical form, which she actually uses if she thinks the hearer will find out the ground function himself. If she does not think so, she will add more material until she thinks she will be understood." (Lestrade 2010: 146)

It is plausible to assume that if the Ground function is exercised by a TOPO speakers can be relatively sure that no extra marking is required for the spatial relation because TOPOs are Grounds par excellence and thus their role in the spatial situation is easily retrievable for the hearer. From the speaker's perspective, the most economical way of coding consists in not coding anything in the first place. COMMs, on the other hand, are not per se Grounds. They either have to be established as Grounds or special communicative conditions must apply to preclude misunderstanding in the absence of material coding. Superficially, this means that TOPOs are always compatible with zero-marking of spatial relations whereas COMMs and ANTHs allow for zero-marking only if the communicative context is favourable to the omission of dedicated markers.

A case in point is the Oceanic language Kaulong about which Ross (2011b: 392) says that

> [p]roper nouns are preceded by a proper noun marker and fall into three subclasses: human masculine (preceded by *a*), human feminine (preceded by *e*), and place names (also preceded by *e*). Human nouns occur after the prepositions *ta* and *hang*, whereas a place name may be used as a locative phrase without a preposition.

In the absence of (in)definite articles, COMMs lack any overt marker which identifies them as members of a given "gender" (Ross 2011b: 393). If we go by the preposed marker alone, TOPOs would fall into the same class as female ANTHs. However, in contrast to the latter and COMMs, TOPOs do not require the presence of a preposition as shown in (126).

(126) Kaulong [Austronesian, Oceanic]
    (a) COMM (Ross 2011b: 405)
        Hiang ni sun [epo pali]$_{PP}$
        3SG.M kill 3.SW **[PREP** spear]
        'He killed him [**with** a spear].' [O.G. and O.T.]
    (b) ANTHs (Ross 2011: 405)
        A Susupa kum [ta e Kristin]
        ANTH.M Susupa work **[PREP** ANTH.F Kristin]$_{PP}$
        'Susupa worked [**for** Christine].' [O.T.]
    (c) TOPO (Ross 2011b: 406)
        Po hek-val [Ø e Au]$_{ADV}$
        3PL quarrel-REC [Ø TOPO Au]
        'They are quarrelling [**at** Au].'

Posture and directional verbs (Ross 2011b: 401) such as *in* 'be at' and *li* 'go to' may govern the Ground-NP directly, i.e., without intervening preposition independent of the word-class (Ross 2011b: 406). There is thus a syntactic context in which COMMs, ANTHs, and TOPOs are equally entitled to zero-marking. In other contexts, however, only TOPOs are privileged in this way.

In (127), we encounter a strategy that resembles that discussed in Section 1.1 in the context of the use of deictics to mark spatial relations in Iloko because "[a] location phrase may consist of or be introduced by a deictic adverb" (Ross 2011b: 407).

(127) Kaulong [Austronesian, Oceanic] (Ross 2011b: 407)
      Po kuk mhang [ili e Nakap]
      3PL be_at house **[up_there** TOPO Nakap]
      'They are in a house [**up** at Nakap].'

Whether this construction type would also tolerate a COMM or an ANTH in the Ground-slot cannot be determined on the basis of the information given in the grammatical sketch of Kaulong. What transpires nevertheless is that different kinds of TOPOs may behave differently in morphosyntax. Except (128), all examples of TOPOs given in Ross (2011b) involve settlement names. In (128), we are dealing with a country name.

(128) Kaulong [Austronesian, Oceanic] (Ross 2011b: 400)
      Minan tin kuli nga-li [Ø Amerika]$_{ADV}$ *men*
      when DEM FUT 1SG.RLS-go_to [Ø America] DELIM
      tapu ku hun kut vamen
      grandmother.POSS.1SG IRR die CMPL DUB
      'When I return [**to** America], my grandmother will possibly have died.'

The Ground *Amerika* is special not because of the absence of a preposition as that is what we expect from a TOPO in this function. Moreover, the motion verb *li* 'go to' forms part of the above-mentioned class of directional verbs which govern the Ground-NP directly without preposition. It is the absence of the TOPO-marker *e* that catches the eye in (128). The absent TOPO-marker invites two different explanations. Either countries differ generally from settlements with the former taking no TOPO-marker whereas it is mandatory with settlement names or the TOPO-marker is obligatorily used only with names of geo-objects which belong to the traditional sphere of experience of the Kaulong speakers. ANTHs do not seem to omit their markers although the data are too scarce to draw any definite conclusions.

Kaulong illustrates the common garden variety of zero-marking of spatial relations which affects first and foremost TOPOs as Ground under the condition that general location is expressed. In the domain of specific location, TOPOs usually require overt marking of the spatial relation in analogy to what is the rule for COMMs. The wide cross-linguistic distribution of zero-marked TOPO-Grounds is amply documented (Stolz et al. 2014, Stolz and Levkovych 2019a, Stolz et al. 2017b) so that there is no need to empirically prove the predilection of TOPOs to remain bare when fulfilling their prototypical function, i.e., the function of Ground. Stolz et al. (2014: 291) conclude that

> [if]f in a given language zero-marking of spatial relations applies, it almost always applies to toponyms.

This cautiously formulated cross-linguistic preference can tentatively be reformulated in a more pointed way as an implicational pattern in Figure 42.

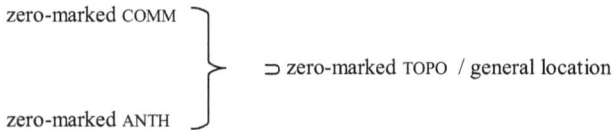

Figure 42: Implicational pattern for zero-marking.

The tenability of this pattern must stand the test in future research dedicated to STG.

Similarly, certain SAG-phenomena call for further scrutiny in the years to come. For the time being, it suffices to present another case of zero-marked ANTH-possessors. In the Oceanic language Iaai, direct possession (aka inalienable possession) is expressed differently according to the word-class of the possessor. As results from (129), COMM-possessors require additional pertensive marking on the possessee whereas ANTH-possessors and their possessees are asyndetically combined.

(129) Iaai [Austronesian, Oceanic] (Lynch 2011: 781)
    (a) COMM-possessor
        *caa-**n***   *laulau*
        leg-**3SG** table
        'the leg of the table' [O.G. and O.T.]
    (b) ANTH-possessor
        caa   *Poou*
        leg   Poou
        'Poou's leg' [O.G. and O.T.]

The genitive relation is thus zero-marked if the possessor is an ANTH and inalienability applies. In the domain of indirect possession (= alienable possession), ANTHS behave like human nouns in general (Lynch 2011: 782). Like in the case of the predictable Ground-function of TOPOS, the possessor role of ANTHS is what one expects in a possessive situation as we have argued throughout Section 4. Interestingly, the predictability of the possessor-hood of ANTHS does not seem to trigger zero-marking as often as the predictability of Ground-hood triggers zero-marking with TOPOS. Whether this impression is borne out by the fact is another of those questions which must be settled in the course of the TYPTOP-project. Figure 43 summarises the findings for Iaai SAG in the usual format of a split rule. Nothing can be said about the behaviour of TOPOS because the grammatical sketch of Iaai does not provide any suitable examples.

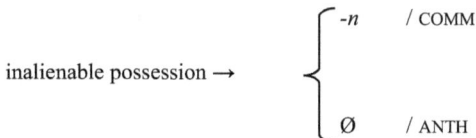

Figure 43: Split rule for inalienable possession in Iaai.

The above examples of zero-marking of TOPOS and ANTHS illustrate the use of representatives of the two PROP-classes as syntactic adjuncts or peripheral participants in Croft's (2022: 173–174) parlance. Is there evidence also for SOG in the domain of (core) arguments or central participants? The answer is yes for ANTHS. With reference to the situation in Roviana, Corston-Oliver (2011: 491) states that

> [p]roper noun phrases occurring as absolutive in main clauses and complement clauses are preceded by the absolutive article *se*, whereas those functioning as ergative are zero marked.

In (130), we reproduce Corston-Oliver's examples to substantiate the claim made in the above quote from the same source.

(130) Roviana [Austronesian, Oceanic] (Corston-Oliver 2011: 491)
  (a) ergative-absolutive pattern
      Seke-i-a   [(e)   Zima]$_{AGENT}$ [se   Maepeza]$_{PATIENT}$
      hit-TRANS-3SG   [(ANTH)   Zima]   [ANTH.ABS   Maepeza]
      'Zima hit Maepeza.' [O.T.]
  (b) S-argument with intransitive verb
      Taloa   [se   Zima]$_{AGENT}$
      leave   [ANTH.ABS   Zima]
      'Zima left.' [O.T.]

Corston-Oliver (2011: 491) emphasises that the optional *e* is not the ergative case-marker but the ANTH-marker which can be used in all syntactic contexts where absolutive case-marking is impossible. The absolutive marker *se* cannot be omitted. Roviana displays split ergativity with ANTHs in main clauses following the ergative system and COMMs the neutral system (Corston-Oliver 2011: 489). The definite article for COMMs is *sa* with different forms for focus, enumeration (= quantification), and plural but not for case except under quantification (Corston-Oliver 2011: 473). In (131) we see that all COMMs are accompanied by *sa* independent of their semantic role.

(131) Roviana [Austronesian, Oceanic] (Corston-Oliver 2011: 489)
      Hiko   pan-i-a   [sa   tie]$_{AGENT}$ [sa   buka]$_{PATIENT}$
      steal   take-TRANS-3SG   [DEF   man]   [DEF   book]
      [koa   sa   barikaleqe]$_{OBLIQUE}$
      [DIR   DEF   woman]
      '[The man] stole [the book] [from the woman].' [O.T.]

In combination with quantifiers (including cardinal numerals), COMMs are also treated according to the ergative-absolutive pattern as shown in (132).

(132) Roviana [Austronesian, Oceanic] (Corston-Oliver 2011: 490)
  (a) S-argument with intransitive verb
      turu   [si   karua   koburu]$_{AGENT}$
      stand   [ABS   two   child]
      '[Two children] are standing up.' [O.G. and O.T.]
  (b) patient in transitive clause
      meke   dogor-i-a   ri$_{AGENT}$ [si   keke   igana]$_{PATIENT}$
      and   see-TRANS-3SG   3PL   [ABS   one   fish]
      '[. . .] and they saw [a fish].' [O.G. and O.T.]

(c) agent in transitive clause
 seke-a [Ø *karua tie*]_AGENT [*sa siki*]_PATIENT
 hit-3SG [Ø two man] [DEF dog]
 '[Two men] hit [the dog].' [O.T.]

The S-argument and the patient are preceded by *si* in (132a–b) whereas the agent in (132c) is zero-marked. It is worth noting that the quantified NPs are translated as indefinites. From the description provided by our source, it cannot be deduced how definite quantified NPs behave.

Corston-Oliver (2011: 491) concludes that

> [s]ince *e* does not mark ergative case, but only the fact that the noun phrase is a proper noun phrase with a human referent, the fact that a noun phrase is [AGENT] must be inferred from the absence of any other indication of case.

On account of the examples in (132), we put forward the following hypothesis. If *e* is the ANTH-marker outside the domain of the absolutive and the absolutive markers of ANTHs and quantified COMMs have an initial sibilant /s/, it is possible to assume that this single consonant is the general absolutive marker so that, at least diachronically, *se* can be decomposed as *s-e* {ABS}-{ANTH}. What to make of the vowels /i/ and /a/ involved in the COMM-article *sa* and the absolutive marker of quantified comms *si* is a question which cannot be answered in this study.

The ergative as such is zero-marked independent of the word-class to which the agent belongs. Unsurprisingly, the examples involve human agents but not exclusively ANTHs. Superficially, this fact seems to speak against taking the Roviana case as proof of ANTHs being zero-marked for a role which is prototypically fulfilled by humans, viz. agent (Croft 2022: 174). Formally, what is zero-marked with ANTHs in the ergative is their being ANTHs. In contrast, their membership in this class is always overtly marked in the absolutive. This distribution of zero-marking and overt marking can be explained functionally. Looking at the situation from the perspective of the ergative, what we expect is to find a human agent in this role. ANTHs are prototypically human and thus need not be extra-marked for this property. This explains why the use of the ANTH-marker *e* is optional in the ergative. In contrast, Croft (2022: 174) argues that patients are most frequently inanimate objects so that human patients are unexpected. Therefore, the ANTH-marker *e* (as part of the absolutive marker *se*) is obligatory for ANTH-patients. In conformity with the ergative-absolutive schema, the marking strategy for patients is the same for S-arguments of intransitive verbs. The latter is enforced by the formal principles of the system whereas the employment of the ANTH-marker in the patient role is connected to the low degree of predictability and expectedness.

In Gela, one of Roviana's many relatives, there is a basic distinction of ANTH(-N)S and COMMS reflected by the opposition of the so-called proper article *a* and the common article *na*. The class of ANTH(-N)S comprises ANTHS, (directly suffixed) kin-terms, and the interrogative *zei* 'who'. Crowley (2011: 527) adds that the members of this class

> are preceded by the article /a/ in subject position. In object position, personal names are obligatorily preceded by the article /a/, while the article is optional before suffixed nouns.

This means that the ANTH(-N)-class can be further divided into ANTHS and ANTH-NS as transpires from the examples in (133).

(133)   Gela [Austronesian, Oceanic]
    (a)   ANTH-subject (Crowley 2011: 527)
        E   inu   beti   [a   Stephen]$_{SUBJECT}$
        3SG.PST drink water [ANTH Stephen]
        'Stephen drank water.' [O.T.]
    (b)   kin-subject (Crowley 2011: 528)
        E   inu   beti   [a   tama-gu]$_{SUBJECT}$
        3SG.PST drink water [ANTH father-1SG]
        'My father drank water.' [O.T.]
    (c)   COMM-subject (Crowley 2011: 528)
        E   inu-vi-a   [na   beti]$_{OBJECT}$ [na   kau]$_{SUBJECT}$
        3SG.PST drink-TRANS-3SG [DEF water] [DEF dog]
        '[The dog] drank [the water].' [O.T.]
    (d)   kin-object (Crowley 2011: 528)
        Geva   ku   riyi-a   [(a)   tama-gu]$_{OBJECT}$
        later 1SG.FUT see-3SG [(ANTH) father-1SG]
        'Later I will see [my father].' [O.T.]

Our source does not provide an example of an ANTH being used as object. This is why we have to take the above quote at face value. Given that ANTHs always require the presence of the ANTH-marker whereas ANTH-Ns may do without it in object function, the possibility of zero-marking of kin-terms is inexplicable if we want to continue with the line of argumentation which rests on predictability and expectedness. Cases which escape to be straightforwardly accounted for by way of applying the most basic functional explanations occur far too often to be dismissed as negligible side-issues. It is a major task of TYPTOP to address instances of SOG which do not visibly conform to the functional-typological expectations. For the time being, we make do with postulating the split rule in Figure 44.

$$\text{article}_{\text{OBJECT}} \rightarrow \begin{cases} a & / \text{ ANTH} \\ (a) & / \text{ kin} \\ na & / \text{ COMM} \end{cases}$$

Figure 44: Split rule for articles in Gela.

A comparable situation is reported for the Salishan language Halkomelem in Canada. We refer to the Musqueam variety of Halkomelem. Halkomelem has a system of articles distinguishing two genders (non-feminine vs feminine) and three deictic categories (proximate, medial, distal) (Suttles 2004: 340). Becker (2021: 314, footnote 13) excludes the so-called oblique article from her account of the Halkomelem system because it "only occurs with proper nouns in oblique position." It is exactly this criterion which makes the case interesting for this study. Suttles (2004: 348) characterises the oblique article $\lambda̓$ as gender-neutralising. Its use is mandatory with PROPs in oblique adjuncts (this also holds for personal pronouns and optionally also for genitival attributes – the latter function being uncommon). The examples in (134) show that ANTHs and TOPOs require the presence of $\lambda̓$.

(134) Halkomelem [Salishan]
    (a) oblique COMM (Suttles 2004: 342)
        ni'   sk^wtéx   [('ə)   tə   léləm̓]_{OBLIQUE}   k^wθə   sq^wəmey̓
        EXIST inside [(OBL) **DEF.M.PROX** house]   DEF.M.DIST dog
        'The dog is [in the house].' [O.T.]
    (b) oblique ANTH (Suttles 2004: 348)
        ćéwatəm   ce'   [('ə) $\lambda̓$ Tom]_{OBLIQUE}   tə   Jack
        be_helped FUT [(OBL) **PROP.OBL** Tom]   DEF.M.PROX Jack
        '(lit.) Jack will be helped [by Tom].' [O.T.]
    (c) oblique TOPO (Suttles 2004: 349)
        'i   [təli'   'ə $\lambda̓$ Seattle]_{PP}
        be_here [from OBL **PROP.OBL** Seattle]
        'He came [from Seattle].' [O.T.]

The free oblique marker 'ə is generally optional (Suttles 2004: 44). This also holds for TOPOs as shown in (135) as opposed to (134c).

(135) Halkomelem [Salishan] (Suttles 2004: 343)
    Ni   ni'   [$\lambda̓$ st^θáməs]_{OBLIQUE}   k^włə   nə-mə́n̓ə
    AUX EXIST [**PROP.OBL** Victoria]   DEF.F.DIST 1SG-child
    'My daughter is [in Victoria].' [O.T.]

The picture arising from the examples can be summarised as Figure 45.

$$\text{oblique} \rightarrow (\textipa{@}?) + \begin{cases} \overset{?}{\lambda} & /\text{ PROP} \\ \\ \text{DEF} & /\text{ COMM} \end{cases}$$

Figure 45: Split rule for oblique case marking in Halkomelem.

The results are contradictory in the sense that on the one hand, the special overt marking of ANTHs as obliques is expected because members of this class are normally expected to fulfil the roles of central participants. They are unpredictable as obliques. On the other hand, TOPOs are prime candidates for occurring in adjuncts and thus function as obliques. Their qualifications notwithstanding, TOPOs are treated on a par with ANTHs so that for once we are facing an instance of Type II (cf. Table 8). What is predictable for ANTHs is not predictable for TOPOs, and vice versa. Unless the motivation behind the distribution of $\overset{?}{\lambda}$ is to mark out PROPs as a class distinct from that of COMMs, the Halkomelem data confirm only to some extent that functional explanations are applicable to the phenomenology of SOG.

The controversial picture painted for Halkomelem notwithstanding, it is clear that not only zero-marking but also overt marking of ANTHs and/or TOPOs in certain syntactic functions can be explained with reference to functional-typological concepts. SOG surfaces frequently in the form of
(a) the absence of dedicated markers in those domains in which the employment of an ANTH or a TOPO is expected, or
(b) by the use of special markers which accompany ANTHs and/or TOPOs in those domains where their presence is not predictable.

In the next sections, we will see whether functional explanations are feasible also with regard to phenomena which do not only involve the presence vs absence of dedicated markers.

6.3 Different ways of expressing the same

SOG is by no means always associated with zero-marking. To the contrary, there is ample evidence of ANTHs and/or TOPOs behaving differently from COMMs as to their choice of constructions which involve the overt expression of a given category. The employment of the Dutch preposition *te* 'in, at' for instance, is characterised as follows by Geerts et al. (1984: 627):

> Het voorzetsel *te* wordt op vier manieren gebruikt. Vóor plaatsnamen is het gelijk aan in; het behoort dan tot de schrijftaal. [...] In tijdsanduidingen is het gelijk aan om; het behorrt dan tot het archaïsch taalgebruik.[143]

Two further contexts in which the use of *te* is common involve infinitival constructions. If we skip the archaic use of *te* in time expressions, it takes TOPOs as its complement when it comes to expressing Place. COMMs and ANTHs are excluded from PPs headed by *te*. To express Place with COMMs, *in* 'in, at' is the first option whereas ANTHs would take *bij* 'at'. In (136), we illustrate this co-existence of options.

(136)   Dutch [Indo-European, Germanic]
    (a) COMM (Geerts et al. 1984: 625)
        [in   de   tram]$_{PP}$ mag je niet roken
        [in   DEF tram]   may 2SG NEG smoke:INF
        '**[In** the tram], you may not smoke.'
    (b) ANTH (Geerts et al. 1984: 625)
        *Hij   logeert   [**bij** Karel]$_{PP}$*
        3SG.M stay:3SG [at   Karel]
        'He stays **[at** Karel's **place]**.'
    (c) TOPO I (Geerts et al. 1984: 627)
        *geboren op   20 september 1935 [**te** 's-Gravenhage]$_{PP}$*
        born   on   20 September 1935 [in 's-Gravenhage]
        'born **[in** 's-Gravenhage] on 20 September, 1935'
    (d) TOPO II (van den Toorn 1977: 237)
        <u>*In*</u> *de vakantie zeilden   we [**in** Friesland]$_{PP}$*
        <u>in</u>   DEF holidays sail:PST:PL 1PL [in Frisia]
        '<u>During</u> the holidays, we sailed **[in** Frisia].'

The distinctions are not entirely neat since depending on style and register *in* can replace *te* in combination with TOPOs as shown in (136d). The preposition *bij* is multifunctional and may combine with COMMs to express relations other than genuinely spatial relations (van den Toorn 1977: 238). It is nevertheless possible to sum up the Dutch givens in the format of the split rule in Figure 46.

---

[143] Our translation: "The preposition *te* is used in four different ways. In front of place names, it is identical to *in*; it belongs to the written language. [...] In time expressions, it is identical to *om*; in these cases, it belongs to the archaic style."

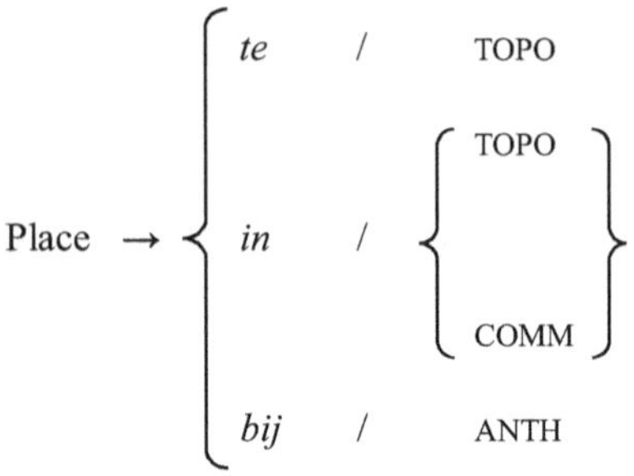

$$\text{Place} \rightarrow \begin{cases} te & / & \text{TOPO} \\ in & / & \begin{Bmatrix} \text{TOPO} \\ \text{COMM} \end{Bmatrix} \\ bij & / & \text{ANTH} \end{cases}$$

Figure 46: Split rule for the expression of Place in Dutch.

Another relatively transparent case is Zulu. Mbeje (2005: 270) claims that "the locative marker is **e-** with names of places, and **kwa-/ka-** with people's names [original boldface]." This wording invokes the formal ANTH ≠ TOPO distinction but leaves open how COMMs behave. In point of fact, TOPOs and COMMs opt for similar but not fully identical strategies as seen in (137).

(137)   Zulu [Atlantic-Congo, Bantu]
    (a) COMM as Ground (Mbeje 2005: 234)
        Ngi-bone   o-Zinhle   n-o-Jabulani   [e-muvi-ni]
        1SG-see   CL1-Zinhle   and-CL1-Jabulani [LOC-movie-LOC]
        'I saw Zinhle and Jabulani [**at** the movies].' [O.T.]
    (b) TOPO as Ground (Mbeje 2005: 257)
        U-Nonhlanhla   u-phesheya   [e-Melika]
        CL1-Nonhlanhla   CL1-overseas [LOC-America]
        'Nonhlanhla is abroad [**in** America].' [O.T.]
    (c) ANTH as Ground (Mbeje 2005: 270)
        Ngi-zo-vakash-ela   [kwa-Zama]   ngo-Khisimuzi
        1SG-FUT-vist-APPL   [LOC.ANTH-Zama]   LOC.CL6-Christmas
        'I'll visit [Zama's **place**] on Christmas.'

As in Swahili (cf. Section 5.5), Zulu makes use of different constructions to express one and the same notion. In (137), three ways of expressing Place are featured. There is a proclitic preposition *kwa-* employed exclusively with ANTH-Grounds for this purpose as shown in (137c). TOPO-Grounds and COMM-Grounds share the proclitic preposition *e-* but the two word-classes differ from each other insofar as TOPOs do not need the suffixal locative marker *-ni* which, however, is obligatory with COMMs. Figure 47 suggests that the absence of zero-marking notwithstanding, there are telling coding asymmetries.

$$\text{place} \rightarrow \begin{cases} \textit{kwa-} & / \text{ANTH} \\ \textit{e-} & / \text{TOPO} \\ \textit{e-} \ldots \textit{-ni} & / \text{COMM} \end{cases}$$

Figure 47: Split rule for Place marking in Zulu.

All three constructions involve overt coding of the spatial relation. However, there are differences as to the segmental and morphological complexity of the expressions. TOPOs stand out insofar as they require only the presence of the monovocalic proclitic *e-* which is the shortest of the three coding strategies. ANTH-Grounds host the segmentally more complex proclitic *kwa-* but are nevertheless expressed in a less complex way than COMM-Grounds since the latter require a bimorphemic combination of the proclitic *e-* and the suffix *-ni*. In canonical morphological terms, ANTHS, TOPOS, and COMMS represent three distinct classes because they express the same differently (Corbett 2007: 30). Moreover, one might want to interpret the above differences in complexity in accordance with the topic of the foregoing Section 6.2 in the sense that the simplest construction is the monopoly of exactly that class which is functionally predestined to fulfil the Ground-function, namely TOPOs. Those classes which are not expected per se to function as Grounds need to be expressly established for this function by extra marking. Situations which resemble that of Zulu however remotely are addressed in the subsequent paragraphs.

The Oceanic language Mussau formally distinguishes two classes of COMMS as well as ANTHS and TOPOS (Ross 2011a: 152). The language is interesting for our project for two reasons. First of all, it provides a parallel to the German adnominal genitive constructions discussed in Section 4.2.1. Ross (2011a: 156) states that

> [i]n the direct possession construction the possessor noun phrase, if any, generally follows the possessed, but may precede it if the possessor is a personal noun.

The examples in (138) show that only ANTH-possessors have the free choice of occupying either the slot to the right or the slot to the left of the possessee whereas COMM-possessors are licit only on the right of the possessee.

(138)   Mussau [Austronesian, Oceanic] (Ross 2011a: 156)
    (a) COMM-possessor
        tama-na   [*aliki*   *eteba*   *toko*]$_{POSSESSOR}$
        father-3SG [child   SG   this]
        '[this child's] father' [O.T.]

(b) ANTH-possessor

tama-na	[Rilu]_{POSSESSOR}	~	[Rilu]_{POSSESSOR}	tama-na
father-3SG	[Rilu]		[Rilu]	father-3SG
'[Rilu's] father'		=	'[Rilu's] father' [O.T.]	

Which of the two options with ANTH-possessors is more frequent or whether there are pragmatic reasons for the choice of the one or the other are questions which cannot be answered on the basis of the data Ross (2011a) provides. These open questions notwithstanding, it is evident that semantically the two linear orders in (138b) are identical. ANTHs are privileged in the sense that speakers can take their pick from two options and exploit their difference for, say, pragmatic purposes whereas COMMs only allow for one construction type.

Outside the domain of possession, the small inventory of prepositions in Mussau is of interest. There are basically only three prepositions, namely *e*, *ta*, and *tei* which are distributed as follows over the word-classes: *e* is responsible for expressing location and time with TOPOs, *ta* encodes location, instrument, and dative with COMMs and ANTHs, and *tei* is the comitative preposition which always takes an object clitic. The preposition *ta* is sensitive to the word-class distinction COMM ≠ ANTH insofar as it takes the form *tale* when it governs a COMM and appears as *ta* when it governs an ANTH (Ross 2011a: 163) as shown in (139).

(139) Mussau [Austronesian; Oceanic] (Ross 2011a: 164)
    (a) COMM as complement
        *Rilu [tale   liu]*_{PP}
        Rilu [LOC:COMM ditch]
        'Rilu is [in the ditch].' [O.T.]
    (b) ANTH as complement
        *A   lao~lao   teke   [ta   Rilu]*_{PP}
        1SG go~PROG there [LOC.ANTH Rilu]
        'I'm going [to Rilu].' [O.T.]
    (c) TOPO as complement
        *A   lao~lao   teke   [e   Palakau]*_{PP}
        1SG go~PROG there [LOC.TOPO Palakau]
        'I'm going [to Palakau village].' [O.T.]

In the light of what we have argued in Section 6.2, the short form *ta* used in combination with ANTHs is remarkable. The long form *tale* is reserved for combinations with COMMs. Is this difference in segmental extension explicable in terms of coding efficiency? In the absence of a suitably large corpus of Mussau, we dare not speculate too wildly. We are nevertheless not convinced that COMMs are ousted quantitatively by ANTHs when it comes to serving as complement of the preposition *ta*. Irrespective of the solution to this problem, a tripartition can be postulated in Figure 48.

$$\text{general location} \rightarrow \begin{cases} e & / \underline{\quad} \text{TOPO} \\ ta & / \underline{\quad} \text{ANTH} \\ tale & / \underline{\quad} \text{COMM} \end{cases}$$

Figure 48: Split rule for general location in Mussau.

The three options in Figure 48 are ordered like those given in Figure 47 for Zulu. The simplest marker (which incidentally also is monovocalic *e*) goes to TOPOs, i.e., to the members of the class of nouny words which are the natural choice when it comes to construing the Ground for a spatial situation. ANTH-Grounds are marked by the slightly more complex CV-preposition *ta* whereas COMM-Grounds require a disyllabic form of the same preposition. The parallel between Zulu and Mussau strongly suggests that there is indeed a language-independent hierarchy of Ground-classes whose internal order calls to mind conceptual differences. This study is not the place, however, to investigate this issue in-depth. This will be the task of further studies to be carried out in the second phase of the project.

The Papuan language Yagaria displays variation in the domain of adnominal possession. As possessors, COMMs and ANTHs boast a plethora of possibilities to express a possessive relationship between nouns, only two of which are shared by the two word-classes (Renck 1975: 169–170). These shared construction types are presented in (140).

(140) Yagaria [Nuclear Trans New Guinea, Goroka] (Renck 1975: 169)
    (a) resumptive pronoun
        ANTH   COMM
        Hane **agae'** bade ≠ Filigano yale **lagae'** aepa
        Hane **POSS.3SG** boy   Filigano people **POSS.1PL** origin
        'Hane's boy'   'the origin **of us** Filigano people' [O.T.]
    (b) pertensive
        ANTH   COMM
        Hane bade-**a'**   bade havú-**'a**
        Hane boy-**POSS.3SG**   boy bow-**POSS.3SG**
        'Hane's boy'   'the boy's bow' [O.T.]

The possessive constructions from which ANTHs are excluded involve the pivotal marker *-ma'* or the nonsingular relation marker *-'i'* suffixed to the possessor as shown in (141).

(141) Yagaria [Nuclear Trans New Guinea, Goroka]
    (a) pivotal marker (Renck 1975: 169)
        *de-**ma'**   bade*
        man-PIV   boy
        'the man's boy' [O.T.]
    (b) relation marker (Renck 1975: 170)
        *yale-**'i'**   yona*
        people-RELL   house
        'the people's house' [O.T.]

Whether the co-existence of the different constructions correlates with any meaning differences or other functional aspects remains unclear. What the constructions in (140)–(141) have in common is the overt marking of the possessive relation. In (142), we present a construction to which zero-marking of the possessive relation applies (the case could have alternatively been discussed already in the previous section).

(142) Yagaria [Nuclear Trans New Guinea, Goroka] (Renck 1975: 169)
    Hane bade
    Hane boy
    'Hane's boy' [O.T.]

The example is synonymous with those for ANTH-possessors in (140a–b). The descriptive grammarian of Yagaria states that juxtaposition of possessor and possessee is possible only if the former is an ANTH and the latter belongs to the class of kinship terms. On account of this restriction, we assume that an alternative translation of (142) could be 'Hane's son'. The possessee represents an inalienable concept. An ANTH can be expected to function as representative of the possessor in a construction which involves a kin-term as possessee whereas a COMM cannot.

It is possible to terminate this section by way of comparing the split rules for inalienable possession in Mussau (cf. (138)) and in Yagaria. Figure 49 accounts for the facts presented above for the Austronesian language. Figure 50, in turn, recapitulates what we have said as to the co-existence of different constructions for the same possessive function in the Papuan language.

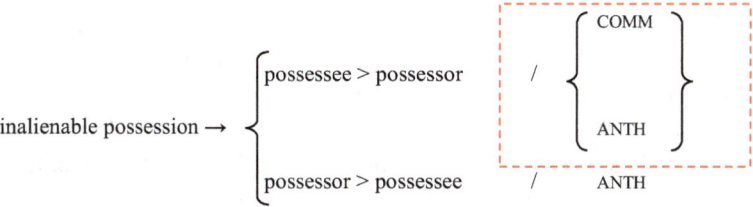

Figure 49: Split rule for inalienable possession in Mussau

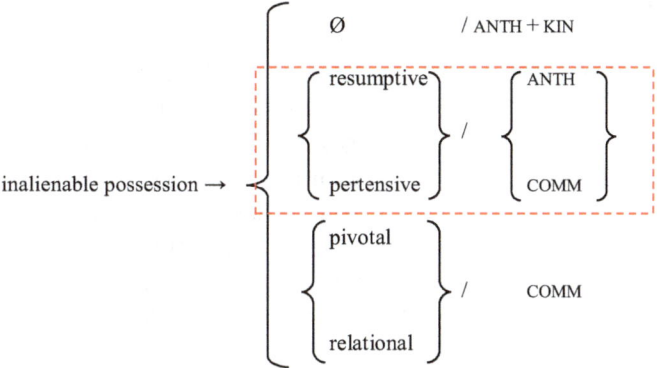

Figure 50: Split rule for inalienable possession in Yagaria.

There are areas of overlap where ANTHs and COMMs are compatible with the same constructions. These overlaps occur in both languages. However, also in both languages, there is always a construction which is reserved for ANTH-possessors. In the case of Mussau, ANTH-possessors alone are entitled to the linear order possessor > possessee. Without further in-depth analysis of the word-order rules in Mussau, it is impossible to decide which of the two possible linearisations is predictable or expected, if at all. In contrast, the richer inventory of possessive constructions attested in Yagaria comprises a case of zero-marking which meets our expectations in the sense that it is possible only for a combination of possessor and possessee which involve an ANTH-possessor.

In spite of the many differences on the micro-level which characterise the cases we have reviewed in this section, it is possible to identify the linking element which connects the one to the other. The languages addressed in Section 6.3 attest to the coexistence of at least two different strategies for expressing a given spatial or possessive relation. The co-existing strategies are functionally synonymous but cannot replace each other because they are licit only in combination with elements which belong to a given word-class. If COMMs are involved, a construction is used that is different from that which is used with ANTHs or TOPOs. Except in the Yagaria case, zero-marking is no option. There is a set of markers which are potential paradigmatic substitutes of each other since they mean the same. Syntagmatically, however, they are restricted to combinations with members of a certain class – combinations from which their synonyms are excluded. In some cases, the exclusiveness is not absolutely strict. Yet, it is always possible to find a niche (if not a larger domain) in which either ANTHs or TOPOs behave according to their own set of rules.

6.4 Mismatchmaking

On the assumption that ANTHs, TOPOs, and COMMs together constitute the macro-class of nouns (Nübling et al. 2015: 28), the following conclusion can be drawn. SOG – be it in the form of SAG or as STG – always challenges the structural homogeneity of this macro-class by way of causing mismatches. According to the logics of Canonical Typology, the (fictitious) canon assumes a set of identical rules on the basis of which all members of the same class behave identically. The pattern COMM = ANTH = TOPO which corresponds to Type I in Table 8 reflects the canon. However, SOG requires special rules which hold only to a segment of the macro-class. Thus, SOG is a source of noncanonicity as captured by the unilateral causal chain in Figure 51.

SOG → noncanonicity

Figure 51: SOG and noncanonicity.

Looking back on the data hitherto presented, we summarily state that each case that was scrutinised is a mismatch no matter how much the behaviour of PROPs and COMMs differ from each other. We have postulated a sizable number of split rules which reflect the impossibility to adequately describe the morphosyntactic properties of PROPs by way of transferring the rules applying to COMMs to PROPs.

For most of the cases studied above, we have explicitly mentioned the kind of mismatch they reflect. A mismatch need not be spectacular. It suffices that a given phenomenon is limited to a small set of PROPs or the other way around, a small set of PROPs is excluded from the application of a certain rule. For several Bantu languages of the Interlacustrine zone (Zone J), Bastin (2003: 521) observes that

> [t]he principal characteristic of Zone J is the [locative prefix] 25, best developed in the western groups JD40, 50 and 60. Its use is limited to proper nouns of place and to 'restrictive locatives'.

In different languages of the zone, this prefix *e-* is attached to TOPOs like *Luganda*, *Musanda*, and *Bushi* to yield the locatives *eluganda* 'at Luganda', *emusanda* 'at Musanda', and *éébushi* 'at Bushi' (Bastin 2003: 521). The mismatch that locative *e-* is responsible for is that of creating inflectional classes because to encode the locative, TOPOs use an affix which differs from the means employed for locative-marking with ANTHs and COMMs. Inflectional classes frequently result from SOG.

The mismatches are not restricted to word-morphology. We have found evidence for SOG also in the interaction of syntactic words. A striking case is reported for the Bantu language Tswana. Gowlett (2003: 627) claims that

> Tswana has a discrete set of genitive concords of the shape *SM-oo-* used to denote communal possession or a larger social unit.

The abbreviation SM can be spelled out as subject marker. For COMM-possessors (except those belonging to class 1a), the genitival concord involves a CV-prefix whose initial consonant is "the typical prevocalic class consonant" (Gowlett 2003: 626). In special genitive constructions, the possessor is always expressed by a PROP (e.g. the name of an individual, a tribal group or a settlement). This phenomenon is probably a case of an informative mismatch (Corbett 2006: 170–171) which applies if the agreement provides "information not available elsewhere."

The data reviewed in this study give evidence of an array of deviations from canonicity. We refer to the following categories in the catalogue of morphological mismatches provided by Corbett (2007: 30):
- **periphrasis**: Lezgian TOPOs (Section 5.1), Latin TOPOs (Section 5.4.2),
- **anti-perphrasis**: Palikur TOPOs (Section 2.4.1), Latin TOPOs (Section 5.4.2),
- **syncretism**: Kryts (Section 4.1), Latin TOPOs (Section 5.4.2),
- **defectiveness**: Vurës TOPOs (Section 2.3.2)
- **overdifferentiation**: Chukchi ANTHs (Section 3) [Estonian COMMs (Section 5.2.3), Iloko COMMs (Section 1.1)]
- **overabundance**: Hungarian TOPOs (Section 5.2.1), Estonian COMMs and TOPOs (Section 5.2.3), Basque TOPOs (Section 5.3), Ancient Greek TOPOs (Section 5.4.1)
- **inflectional classes**: Iloko (Section 1.1), Nungon (Section 2.3.1), Chechen (Section 3), Koasati (Section 3), Kryts (Section 4.1), German (Section 4.2.1), Faroese (Section 4.2.2), Romanian (Section 4.4), Lezgian (Section 5.1), Hungarian (Section 5.2.1), Basque (Section 5.3), Ancient Greek (Section 5.4.1), Latin (Section 5.4.2), Zulu (Section 6.3), Zone J Bantu (Section 6.4)
- **deponency**: Lezgian TOPOs (Section 5.1)

These cases only involve morphological paradigms. There are further mismatches on the morphosyntactic level which need not be recapitulated here since they have been highlighted time and again throughout the empirical parts of this study. We do not claim that the above list is exhaustive as to the types of mismatches one can find in the domain of SOG. Zero-marking is a further candidate for the role of morphological mismatch still missing from Corbett's list (Stolz and Levkovych 2019b). Paradigmatic uninflectability (Nübling 2012) as opposed to syntagmatic uninflectedness (~ zero-marking) (Doleschal 2023) should be included in future studies.

The evidence adduced from languages of different genetic, structural, and geographic background strongly suggests that SOG should be taken account of within the framework of Canonical Typology since ANTHs and TOPOs have been shown to trigger mismatches, some major, some minor ones, and thus contribute to noncanonicity. They are potential mismatchmakers. They assume this role across many languages. SOG-induced mismatches are no marginal entity in the structural phenomenology of human languages. Ignoring them systematically is tantamount to denying them the membership in the macro-class of nouns.

7 Conclusions and outlook

This study has achieved both of its major goals. On the one hand, the plethora of data presented unequivocally proves that SOG is a cross-linguistically recurrent phenomenon that calls for being scrutinised in as much detail as possible by typologists. On the other hand, the two parts of the working hypothesis (27) have been confirmed. SAG has a stronghold in the domain of possession whereas STG is especially well-established in the grammar of space. This fact alone motivates the continuation of linguistic research on the morphosyntax of PROPs in cross-linguistic perspective. Onomasticians can gain from studies of this kind insofar as it is shown that there is more to PROPs than their etymology. Proponents of the Grammar of Names are given the opportunity to systematically investigate the structural and functional similarities and dissimilarities of PROPs across languages. Linguists who inquire into possession or spatial relations might find it worthwhile including ANTHs and TOPOs more prominently in their research. Similarly, Canonical Typologists might want to pay more attention to PROPs because of their potential of causing morphological and other structural mismatches. The expectations are high for a bright future of SOG-related projects conducted by scholars of different theoretical convictions and methodological orientations. However, this speculation about what might happen in times to come does not declare the case closed for us. The agenda of TypTop contains a long list of items which still need to be addressed in dedicated follow-up studies.

It has been shown that functional-typological explanations which take predictability and expectedness as their guidelines are indeed applicable to many cases of coding asymmetry both of the zero-marking kind and the reduced segmental complexity kind. However, SOG cannot be equated with these aspects, at least not always straightforwardly. There are also phenomena which seem to run counter to the usual implications of predictability and expectedness. We admit that some of the finer points of the functional-typological dialects probably escape us but the spatio-deictic clitics exclusively attached to TOPOs in Ancient Greek (Section 5.4.1) and the overt marking of the possessive relation with ANTH-possessors as opposed to simple juxtaposition in the case of COMM-possessors in Amele (Section 4.1) seem to speak against sweepingly generalising efficiency-based explanations. These two cases of more complex expressions for TOPOs and ANTHs exactly in those contexts where they are entitled to zero-marking or reduced segmental complexity in many other languages are examples of inefficiency because the predictable and expected is expressed in a costlier way than those word-classes which are neither predictable nor expected in the function. In Haspelmath (2021: 608) – an up-date of Haspelmath (2019) – the author hedges his claims by saying that when I claim that a grammatical coding asymmetry is universal, I do not mean to exclude the possibility of individual exceptions, and I merely claim that the asymmetry is found with much greater than chance frequency.

Indeed, our own database largely corroborates what Haspelmath (2021: 617–618) observes with reference to differential place marking:

> In a substantial number of languages, locative flagging is differential, such that place names tend to have the shortest coding (often zero, as in Tswana), and animate nouns tend to have the longest coding, with inanimate nouns intermediate between the two [. . .] In Basque, we see a three-way contrast, while Tswana exemplifies a contrast between place names and inanimates, and Tamil shows a contrast between inanimates and animates. Again, this is because place names are usually associated with locative use, while this is less common for inanimate nouns, let alone animate nouns. The parallels with differential object marking are so striking that [I] call[. . .] this pattern 'differential place marking'.

However, there is as yet no cross-linguistic study based on a sizable balanced sample which determines how frequently violations of the coding asymmetry in the spatial domain are attested with TOPOs functioning as Grounds. Chances are that cases like Ancient Greek are indeed (very) infrequent. Yet, the small number of instances in this study which pose problems for the concept of differential place marking as put forward by Haspelmath cannot be taken to be representative of the cross-linguistic distribution of phenomena of this kind since the languages reviewed in the foregoing sections constitute a convenience sample. There is thus a topic for further research in the spirit of TYPTOP.

Cases like that of Halkomelem oblique marking of PROPs and the absence of ANTH-marking in the ergative in Mussau are also interesting in this context because they may serve as extension of Haspelmath's (2021: 611–612 and 617) assumptions about the asymmetrical marking of nominatives and accusatives and his ideas on the ergative. What we see in the SOG-cases is that PROPs are obligatorily marked in the unexpected function (patient or oblique) whereas they are unmarked when the function is predictable for a human participant (subject or ergative). We add that TOPOs are excluded from argument roles in Vurës but still co-occur with their obligatory class marker. Haspelmath (2021: 618–619) mentions different adnominal possessive constructions without, however, looking at cases as those where human possessors are unmarked whereas inanimate possessors are marked as such. The notion of differential possessor marking is already lurking beneath the surface, in a manner of speaking.

We do not only ignore how often STG fails to conform to the functional-typological expectations. There is also a dire need for determining the extent to which SOG in general and SAG and STG in particular are represented in the languages of the world. Again, this goal can be achieved only on the basis of a much larger balanced sample.

At the same time, ticking off those languages which attest to SOG and separating them from which do not is one thing. It is also highly interesting to find out about the text frequencies of the constructions in those cases where there is competition between SOG-constructions and others which are shared with COMMs. This is also an indirect way of testing efficiency if we go by Zipf's Law of Abbreviation which predicts the shrinking size of constructions (originally: words) in correlation with increasing frequency of use (Zipf 1968: 38). In the course of the empirical illustration of SOG, we have repeatedly mentioned the possibility of inefficiency arising in the context of SOG. It remains to be seen whether this contention can be upheld if further data and languages are looked at. The most serious problem with this task is the limited availability of large enough corpora in the sample languages. If this part of the project can be carried out only for a subset of the future sample, it must be explained on which grounds it is possible to generalise the findings over the entire sample and beyond.

We have left a number of questions unanswered which have to be put on the future to-do list of TypTop. A particularly important issue can only be vaguely characterised at this point. There are the Lezgian and Uralic cases which show that TOPOs referring to the traditional historical territory of a speech-community are distinguished morphosyntactically from the bulk of TOPOs, most of which refer to geo-objects outside this narrowly circumscribed area. It is not immediately obvious how and whether this differential treatment of TOPOs can be integrated into an efficiency-based model or into that of differential place marking. Glossing over a number of differences, we loosely associate with these cases the zero-marking of spatial relations (Place and Goal) for familiar hodo- and dromonyms in French (Stolz et al. 2014: 193–195) and the zero-marking of the same categories with familiar settlement names in languages such as Lango (Stolz et al. 2014: 102–103). The cases of zero-marking might be causally connected to predictability and expectedness. But what about the choice of a particular series of spatial cases without any net-gain in terms of reduced segmental chains?

Something we have touched upon only once and in passing is Nübling's (2017a) concept of onymic schema constancy (German: *Namenkörperschonung*). According to this principle, PROPs tend to be immune against phonological and morphological processes which would interfere with their segmental shape. The immunity is functionally important because it guarantees the easy and permanent recognisability of a given PROP. Schema constancy is not only attested for the segmental level. For the Bantu language Herero, Elderkin (2003: 605) states that tone-sandhi which affects COMMs in combination with preceding prepositions is blocked for ANTHs:

> Where the following syllable is high, or **the first syllable of a proper noun (the tones on which may not be perturbed)**, the preposition takes the high tone and, where possible, the low appears on the previous syllable [added boldface].

It can be assumed that the blockage of the application of suprasegmental rules which would alter the prosodic structure of an ANTH or TOPO is found in many tone languages. At the same time, we know from the discussion in this study that clipping and truncating of ANTHs to create vocatives is a cross-linguistically wide-spread practice which seems to be stronger than the principle of schema constancy. There is thus an apparent paradox which should be solved in a dedicated study of its own.

This study is coming to a close. There are many loose ends which still call for being tied together. That there are so many of them is indicative of the fertile ground linguists of different persuasions might find in the Grammar of Names and particularly in SOG. We hope that readers have acquired a taste for doing research in the areas outlined in this study. The TypTop-project has only just started and will certainly continue further. The sample will be considerably extended as will the descriptive linguistic sources. A questionnaire is intended to support us in systematically gathering pertinent data. It is hoped that many of the gaps in our knowledge about SOG can be filled to the benefit not only of language typology but all other disciplines which take an interest in PROPs.

Primary sources

Ale; Afr = C. Iulius Caesar [translated by Anton Baumstark, revised and commented upon by Carolin Jahn]. 2004. *Gaius Iulious Caesar. Kriege in Alexandrien, Afrika, und Spanien. Lateinisch und deutsch.* Darmstadt: Wissenschaftliche Buchgesellschaft.
Bud = Vilmos Kondor. 2012. *Budapest Novemberben.* Budapest: Agave.
Cae = Aurelius Victor [edited, translated and annotated by Kirsten Groß-Albenhausen & Manfred Fuhrmann]. 1997. *S. Aurelius Victor: Die Römischen Kaiser / Liber de Caesaribus. Lateinisch-deutsch.* Darmstadt: Wissenschaftliche Buchgesellschaft.
Gal = C. Iulius Caesar [edited and translated by Marieluise Deißmann]. 1980. *De Bello Gallico / Der Gallische Krieg.* Stuttgart: Reclam.
Dir = Marianna Debes Dahl. 1979. *Dirdri.* Tórshavn: Fannir.
Dun = Muhammad said Abdulla. 1992. *Duniani kuna watu.* Nairobi, Kampala, Dar es Salaam: East African Educational Publishers.
HP I Basque = Iñaki Mendiguren. 2000. *Harry Potter eta Sorgin Harria.* Donostia: Elkarlanean.
HP I Dutch = Wiebe Buddingh. 2001. *Harry Potter en de stehen der wijzen.* Amsterdam, Antwerpen: De Harmonie, Standaard.
HP I English = Joanne K. Rowling. 2000. *Harry Potter and the phlosopher's stone.* London: Bloomsbury.
HP I Estonian = Krista Kaer & Kaisa Kaer. 2005. *Harry Potter ja tarkade kivi.* Tallinn: Varrak.
HP I Faroese = Gunnar Hoydal. 2000. *Harry Potter og vitramannasteinurin.* Tórshavn: Bókadeild Føroya Lærafelags.
HP I Finnish = Jaana Kapari-Jatta. 1998. *Harry Potter ja viisasten kivi.* Helsinki: Tammi.
HP I Hungarian = Tóth Tamás Boldizsár. 2001. *Harry Potter és a bölcsek köve.* Budapest: Animus.
HP I Icelandic = Helga Haraldsdóttir. 1999. *Harry Potter og viskusteinninn.* Reykjavík: Bjartur.
HP I Low German = Hartmut Cyriacks & Peter Nissen. 2002. *Harry Potter un de Wunnersteen.* Kiel: Jung.
HP I Romanian = Ioana Iepureanu. 2000. *Harry Potter şi piatra filozofală.* Bucureşti: Egmont.
HP I West Frisian = Jetske Bilker. 2007. *Harry Potter en de stien fan 'e wizen.* Ljouwert, Utert: Bornmeer.
Jón = Guðrún Helgadóttir. 1995. *Jón Oddur og Jón Bjarni.* Reykjavík: Vaka-Helgafell.
LPP Swahili = Philipp Kruse & Walter Bgoya. 2009. *Mwana mdogo wa mfalme.* Dar es Salaam: Mkuki na Nyota Publishers.
Odu = Homer [edited by Roland hampe]. 2010. *Odyssee. Griechisch/deutsch.* Stuttgart: Reclam.
Reg = Heini Hestoy. 1991. *Regin og Sára.* Tórshavn: Føroya Lærafelag.
Róz = G. Szábo Judit. 2005. *Rózsás Letícia.* Budapest: Móra.
Sum = Heini Hestoy. 1995. *Summardáar.* Tórshavn: Bókadeild Føroya Lærafelags.
Táv = Mary Westmacott [aka Agatha Christie] [translated by Livia Görög]. 2005. *Távol telt tőled tavaszom.* Budapest: Partvonal.

References

Academia de Lenguas Mayas de Guatemala. 2006. *Runuk'ulem pa rub'eyal rutz'ib'axik ri Kaqchikel ch'ab'äl. Grmática normative del idioma Kaqchikel*. Iximulew, Chimaltenango: Instituto de Lingüística Universidad Rafael Landivar de Guatemala.
Ackermann, Tanja. 2018. From genitive inflection to possessive marker? In Tanja Ackermann, Horst Simon & Christian Zimmer (eds.), *Germanic genitives*, 189–232. Amsterdam, Philadelphia: Benjamins.
Ackermann, Tanja, Horst Simon & Christian Zimmer (eds.). 2018. *Germanic genitives*. Amsterdam, Philadelphia: Benjamins.
Adelaar, Willem F.H. 2017. Imperatives and commands in Quechua. In Alexandra Y. Aikhenvald & R.M.W. Dixon (eds.), *Commands. A cross-linguistic typology*, 46–60. Oxford: Oxford University Press.
Aikhenvald, Alexandra Y. 2000. *Classifiers. A Typology of Noun Categorization Devices*. Oxford: Oxford University Press.
Aikhenvald, Alexandra Y. 2010. *Imperatives and commands*. Oxford: Oxford University Press.
Aikhenvald, Alexandra Y. 2013. Possession and ownership. A cross-linguistic perspective. In Alexandra Y. Aikhenvald & R.M.W. Dixon (eds.), *Possession and ownership. A cross-linguistic typology*, 1–64. Oxford: Oxford University Press.
Aikhenvald, Alexandra Y. 2015. *The art of grammar. A practical guide*. Oxford: Oxford University Press.
Ainiala, Terhi. 2016. Names in society. In Carole Hough (ed.), *The Oxford handbook of names and naming*, 371–381. Oxford: Oxford University Press.
Aldrin, Emilia. 2016. Names and identity. In Carole Hough (ed.), *The Oxford handbook of names and naming*, 382–394. Oxford: Oxford University Press.
Allerton, D.J. 1996. Proper names and definite descriptions with the same reference. A pragmatic choice for language users. *Journal of Pragmatics* 25(5). 621–633.
Anderson, John M. 1991. Notional grammar and the redundancy of syntax. *Studies in Language* 15(2). 301–333.
Anderson, John M. 2004. On the grammatical status of names. *Language* 80(3). 435–474.
Anderson, John M. 2007. *The grammar of names*. Oxford: Oxford University Press.
Authier, Gilles. 2009. *Grammaire kryz (langue caucasique d'Azerbaïdjan, dialecte d'Alik)*. Leuven, Paris: Peeters.
Baerman, Matthew. 2007. Morphological typology of deponency. In Matthew Baerman, Greville G. Corbett, Dunstan Brown & Andrew Hippisley (eds.), *Deponency and morphological mismatches*, 1–20. Oxford: Oxford University Press.
Baerman, Matthew & Greville G. Corbett. 2010. Introduction. Defectiveness: Typology and diachrony. In Matthew Baerman, Greville G. Corbett & Dunstan Brown (eds.), *Defective paradigms. Missing forms and what the tell us*, 1–18. Oxford: Oxford University Press.
Bánhidi, Zoltán, Zoltán Jókay & Dénes Szabó. 1975. *Lehrbuch der ungarischen Sprache*. München: Hueber.
Bárczi, Géza. 2001. *Geschichte der ungarischen Sprache*. Innsbruck: Universität Innsbruck.
Bastin, Yvonne. 2003. The Interlacustrine zone (Zone J). In Derek Nurse & Gérard Philippson (eds.), *The Bantu languages*, 501–528. London, New York: Routledge.
Bauer, Gerhard. 1995. Namenforschung im Verhältnis zu anderen Forschungsdisziplinen. In Ernst Eichler, Gerold Hilty, Heinrich Löffler, Hugo Steger & Ladislav Zgusta (eds.), *Namenforschung*. 1. Halbband, 8–23. Berlin: De Gruyter.
Bauer, Winifred. 1993. *Maori*. London, New York: Routledge.

Becker, Laura. 2021. *Articles in the world's languages*. Berlin, Boston: De Gruyter.
Bendel, Christiane. 2006. *Baskische Grammatik*. Hamburg: Buske.
Berchtold, Simone & Antje Dammel. 2014. Kombinatorik von Artikel, Ruf- und Familiennamen in Varietäten des Deutschen. In Friedhelm Debus, Rita Heuser & Damaris Nübling (eds.), *Linguistik der Familiennamen*, 249–280. Hildesheim: Olms.
Berghäll, Liisa. 2015. *A grammar of Mauwake*. Berlin: Language Science Press.
Beyrer, Arthur, Klaus Bochmann & Siegfried Bronsert. 1987. *Grammatik der rumänischen Sprache der Gegenwart*. Leipzig: Enzyklopädie.
Blake, Barry J. 1994. *Case*. Cambridge: Cambridge University Press.
Boisserie, Étienne. 2008. Aspects historiques et juridiques des question toponymiques dans le sud de la Slovaquie. In Hervé Guillorel (éd.), *Toponymie et politique. Les marqueurs linguistiques du territoire*, 117–126. Bruxelles: Brulyant.
Booij, Geert. 2002. *The morphology of Dutch*. Oxford: Oxford University Press.
Booij, Geert. 2013. Morphology in construction grammar. In Thomas Hoffmann & Graeme Trousdale (eds.), *The Oxford handbook of construction grammar*, 255–273. Oxford: Oxford University Press.
Bornemann, Eduard & Ernst Risch. 1978. *Griechische Grammatik*. Frankfurt a.M.: Diesterweg.
Bosque, Ignacio (ed.). 2009. *Nueva gramática de la lengua Española. Morfología, sintaxis I*. Madrid: Real Academia Española.
Boyeldieu, Pascal. 2019. Proper names and case markers in Sinyar (Chad/Sudan). *Language Typology and Universals (STUF)* 72 (4). 467–504.
Brauner, Siegmund & Irmtraud Herms. 1986. *Lehrbuch des modernen Swahili*. Leipzig: Enzyklopädie.
Braunmüller, Kurt. 2018. On the role of case and possession in Germanic. A typological approach. In Tanja Ackermann, Horst Simon & Christian Zimmer (eds.), *Germanic genitives*, 301–323. Amsterdam, Philadelphia: Benjamins.
Broderick, George. 2010. Manx. In Martin J. Ball & Nicole Müller (eds.), *The Celtic languages*, 305–356. London, New York: Routledge.
Brown, Dunstan & Marina Chumakina. 2013. What there might be and what there is. An introduction to Canonical Typology. In Dunstan Brown, Marina Chumakina & Greville G. Corbett (eds.), *Canonical morphology and syntax*, 1–19. Oxford: Oxford University Press.
Brown, Penelope & Stephen C. Levinson. 1987. *Politeness. Some universals in language usage*. Cambridge: Cambridge University Press.
Brown, R. McKenna, Judith M. Maxwell & Walter E. Little. 2006. *¿La ütz awäch? Introduction to Kaqchikel Maya language*. Austin: University of Texas Press.
Caragiu Marioţeanu, Matilda. 1975. *Compendiu de dialectologie română (nord- şi suddunăreană)*. Bucureşti: Editura ştiinţifică şi enciclopedică.
Caro Reina, Javier. 2014. The grammaticalization of the terms of address *en* and *na* as onymic markers in Catalan. In Friedhelm Debus, Rita Heuser & Damaris Nübling (eds.), *Linguistik der Familiennamen*, 175–204. Hildesheim: Olms.
Caro Reina, Javier. 2020a. The definite article with place names in Romance languages. In Nataliya Levkovych & Julia Nintemann (eds.), *Aspects of the grammar of names. Empirical case studies and theoretical topics*, 25–52. München: LINCOM.
Caro Reina, Javier. 2020b. Differential object marking with proper names in Romance languages. In Luise Kempf, Damaris Nübling & Mirjam Schmuck (Hg.), *Linguistik der Eigennamen*, 225–254. Berlin, New York: De Gruyter.
Caro Reina, Javier. 2022. The definite article with personal names in Romance languages. In Javier Caro Reina & Johannes Helmbrecht (eds.), *Proper names versus common nouns. Morphosyntactic contrasts in the languages of the world*, 51–92. Berlin, Boston: De Gruyter Mouton.

Caro Reina, Javier & Johannes Helmbrecht. 2022. Morphosyntactic contrasts between proper names and common nouns. An introduction. In Javier Caro Reina & Johannes Helmbrecht (eds.), *Proper names versus common nouns. Morphosyntactic contrasts in the languages of the world*, 1–20. Berlin, Boston: De Gruyter Mouton.

Caro Reina, Javier & Jessica Nowak. 2019. Diachronic development of gender in city names in Spanish. *Language Typology and Universals (STUF)* 72 (4). 505–538.

Chantraine, Pierre. 1963. *Grammaire Homérique. Tome II: Syntaxe*. Paris: Klincksieck.

Chappell, Hilary & William McGregor (eds.). 1995. *The grammar of inalienability. A typological perspective on body part terms and part-whole relations*. Berlin, New York: Mouton de Gruyter.

Cojocaru, Dana. 2004. *Romanian grammar*. Durham/NC: Duke University.

Comrie, Bernard. 1986. Markedness, grammar, people, and the world. In Fred R. Eckman, Edith A. Moravcsik & Jessica Wirth (eds.), *Markedness*, 85–196. New York: Plenum Press.

Comrie, Bernard & Norval Smith. 1977. The Lingua descriptive studies questionnaire. *Lingua* 42(1). 1–72.

Corbett, Greville G. 1991. *Gender*. Cambridge: University of Cambridge Press.

Corbett, Greville G. 2006. *Agreement*. Cambridge: Cambridge University Press.

Corbett, Greville G. 2007. Deponency, syncretism, and what lies in between. In Matthew Baerman, Greville G. Corbett, Dunstan Brown & Andrew Hippisley (eds.), *Deponency and morphological mismatches*, 21–44. Oxford: Oxford University Press.

Corbett, Greville G. 2013. Canonical morphosyntactic features. In Dunstan Brown, Marina Chumakina & Greville G. Corbett (eds.), *Canonical morphology and syntax*, 48–65. Oxford: Oxford University Press.

Cornilescu Alexandra & Alexandru Nicolae. 2015. Classified proper names in Old Romanian. Person and definiteness. In Virginia Hill (ed.), *Formal approaches to DPs in Old Romanian*, 100–153. Leiden, Boston: Brill.

Corston-Oliver, Simon. 2011. Roviana. In John Lynch, Malcolm Ross & Terry Crowley (eds.), *The Oceanic languages*, 467–497. London, New York: Routledge.

Creissels, Denis. 2006. *Syntaxe Générale. Une introduction typologique I: Catégories et constructions*. Paris: Lavoisier.

Croft, William. 2001. *Radical construction grammar. Syntactic theory in typological perspective*. Oxford: Oxford University Press.

Croft, William. 2022. *Morphosyntax. Constructions of the world's languages*. Cambridge: Cambridge University Press.

Crowley, Terry. 2011. Gela. In John Lynch, Malcolm Ross & Terry Crowley (eds.), *The Oceanic languages*, 525–537. London, New York: Routledge.

Cysouw, Michael & Jeff Good. 2013. Languoid, Doculect, Glossonym. Formalizing the notion 'language'. *Language Documentation & Conservation* 7. 331–360.

Cysouw, Michael & Jan Wohlgemuth. 2010. The other end of universals: theory and typology of rara. In Jan Wohlgemuth & Michael Cysouw (eds.), *Rethinking universals. How rarities affect linguistic theory*, 1–10. Berlin, Boston: De Gruyter Mouton.

Dahl, Östen & Maria Koptjevskaja-Tamm. 2001. Kinship in grammar. In Irène Baron, Michael Herslund & Finn Sørensen (eds.), *Dimensions of possession*, 201–226. Amsterdam, Phildaelphia: Benjamins.

Daniel, Michael & Edith Moravcsik. 2005. The associative plural. In Martin Haspelmath, Matthew S. Dryer, David Gil & Bernard Comrie (eds.), *The world atlas of language structures*, 150–153. Oxford: Oxford University Press.

Daniel, Michael & Andrew Spencer. 2009. The vocative – an outlier case. In Andrej Malchukov & Andrew Spencer (eds.), *The Oxford handbook of case*, 626–634. Oxford: Oxford University Press.

Davies, Jon. 1989. *Kobon*. London, New York: Routledge.
De Rijk, Rudolf P.G. 2008. *Standard Basque. A Progressive Grammar.* Cambridge/Mass.: MIT.
Delsing, Lars-Olof. 1993. *The internal structure of noun phrases in the Scandinavian languages. A comparative study*. Lund: Lunds Universitet.
D'hulst, Yves, Rolf Thieroff & Trudel Meisenburg. 2022. River names. Definite articles and place names in West-Germanic and Romance. In Javier Caro Reina & Johannes Helmbrecht (eds.), *Proper names versus common nouns. Morphosyntactic contrasts in the languages of the world*, 93–120. Berlin, Boston: De Gruyter Mouton.
Dimitrescu, Florica, Viorica Pamfil, Elena Barborică, Maria Cvasnîi, Mirela Theodorescu, Cristina Călăraşu, Mihai Marta, Elena Toma & Liliana Ruxăndoiu. 1978. *Istoria limbii române*. Bucureşti: Editura didactică şi pedagogică.
Dixon, R.M.W. 2010a. *Basic linguistic theory. Volume 1: Methodology*. Oxford: Oxford University Press.
Dixon, R.M.W. 2010b. *Basic linguistic theory. Volume 2: Grammatical topics*. Oxford: Oxford University Press.
Doleschal, Ursula. 2023. *The conditions of uninflectability in nouns in the Slavic languages*. Talk delivered on occasion of the 45th *Annual Conference of the Deutsche Gesellschaft für Sprachwissenschaft (DGfS)* in Cologne (7–10 March, 2023).
Draskau, Jennifer Kewley. 2008. *Practical Manx*. Liverpool: University of Liverpool Press.
Ebert, Karen H. 1979. *Sprache und Tradition der Kera (Tschad). Teil III: Grammatik*. Berlin: Reimer.
Eichler, Ernst. 1995. Entwicklung der Namenforschung. In Ernst Eichler, Gerold Hilty, Heinrich Löffler, Hugo Steger & Ladislav Zgusta (eds.), *Namenforschung*. 1. Halbband, 1–7. Berlin: De Gruyter.
Eichler, Ernst, Gerold Hilty, Heinrich Löffler, Hugo Steger & Ladislav Zgusta (eds.). 1995. *Namenforschung*. 1. Halbband. Berlin: De Gruyter.
Eisenbeiß, Sonja, Ayumi Matsuo & Ingrid Sonnenstuhl. 2009. Learning to encode possession. In William B. McGregor (ed.), *The expression of possession*, 143–212. Berlin, New York: Mouton de Gruyter.
Elderkin, Edward D. 2003. Herero. In Derek Nurse & Gérard Philippson (eds.), *The Bantu languages*, 581–608. London, New York: Routledge.
Enfield, N.J. 2017. Linguistic expression of commands in Lao. In Alexandra Y. Aikhenvald & R.M.W. Dixon (eds.), *Commands. A cross-linguistic typology*, 189–205. Oxford: Oxford University Press.
Erelt, Tiiu (ed.). 2002. *Eesti keele sõnaraamat ÕS 1999*. Tallinn: Eesti Keele Sihtasutus.
Espiritu, Precy. 1984. *Let's speak Ilokano*. Honolulu: University of Hawaii Press.
Fedden, Sebastian. 2011. *A grammar of Mian*. Berlin, Boston: De Gruyter Mouton.
Flores, P. F. Ildefonso Joseph. 1753. *Arte de la lengua metropolitana del Reyno Cakchiquel, o Gvatemalico con un parallel de las lenguas metropolitanas de los Reynos Kiche, Cakchiquel, y ₄ₜutuhil, que hoy integran el Reyno de Guatemala*. Guatemala: Sebastián de Arebalo.
Forker, Diana. 2020. *A grammar of Sanzhi Dargwa*. Berlin: Language Science Press.
Fried, Mirjam. 2009. Plain vs situated possession in Czech. A constructional account. In William B. McGregor (ed.), *The expression of possession*, 213–249. Berlin, New York: Mouton de Gruyter.
Geerts, G., W. Haeseryn, J. de Rooij & M.C. van den Toorn. 1984. *Algemene Nederlandse spraakkunst*. Groning, Leuven: Wolters-Noordhoff.
Gemoll, Wilhelm. 1979. *Griechisch-deutsches Schul- und Handwörterbuch*. München, Wien: Freytag & Hölder-Pichler-Tempsky. [1st edition 1908]
Gildersleeve, B.L. & Gonzalez Lodge. 1968. *Latin Grammar*. 3rd edn. London: Macmillan.
Givón, T. 2017. *The story of zero*. Amsterdam, Philadelphia: Benjamins.
Goldap, Christel. 1991. *Lokale Relationen im Yukatekischen*. Frankfurt a.M., Bern: Lang.
Gönczöl, Ramona. 2020. *Romanian. An essential grammar*. London, New York: Routledge.

Gowlett, Derek. 2003. Zone S. In Derek Nurse & Gérard Philippson (eds.), *The Bantu languages*, 609–638. London, New York: Routledge.
Guzmán Naranjo, Matías & Laura Becker. 2021. Coding efficiency in nominal inflection. Expectedness and type frequency effects. *Linguistic Vanguard* 7(s3), 20190075.
Handschuh, Corinna. 2017. Nominal category marking on personal names. A typological study of case and definiteness. *Folia Linguistica* 51(2). 483–504.
Handschuh, Corinna. 2019. The classification of names. *Language Typology and Universals (STUF)* 72(4). 539–572.
Handschuh, Corinna. 2022. Personal names versus common nouns. Crosslinguistic findings from morphology and syntax. In Javier Caro Reina & Johannes Helmbrecht (eds.), *Proper names versus common nouns. Morphosyntactic contrasts in the languages of the world*, 21–50. Berlin, Boston: De Gruyter Mouton.
Handschuh, Corinna & Antje Dammel. 2019. Introduction: Grammar of names and grammar out of names. *Language Typology and Universals (STUF)* 72(4). 453–466.
Harries, L. 1977. The syntax of Swahili locative affixes. *African Studies* 36(2). 171–186.
Harweg, Roland. 1999. *Studien zu Eigennamen*. Aachen: Shaker.
Haspelmath, Martin. 1993. A Grammar of Lezgian. Berlin, New York: Mouton de Gruyter.
Haspelmath, Martin. 2000. Periphrasis. In Geert Booij, Christian Lehmann & Joachim Mugdan (eds.), *Morphology: A handbook on inflection and word formation*. Vol. 1, 654–664. Berlin: De Gruyter.
Haspelmath, Martin. 2010. Framework-free grammatical theory. In Bernd Heine & Heiko Narrog (eds.), *The Oxford handbook of linguistic analysis*, 287–310. Oxford: Oxford University Press.
Haspelmath, Martin. 2019. Differential place marking and differential object marking. *Language Typology and Universals (STUF)* 72(3). 313–334.
Haspelmath, Martin. 2021. Explaining grammatical coding asymmetries. Form–frequency correspondences and predictability. *Journal of Linguistics* 57. 605–633.
Haspelmath, Martin & Andrea D. Sims. 2010. *Understanding Morphology*. London: Hodder Education.
Heidenkummer, Alexandra & Johannes Helmbrecht. 2017. Form, Funktion und Grammatikalisierung des Eigennamenmarkers =ga im Hoocąk (Sioux). In Johannes Helmbrecht, Damaris Nübling & Barbara Schlücker (eds.), *Namengrammatik*, 11–32. Hamburg: Buske.
Heine, Bernd. 1997. *Possession. Cognitive sources, forces, and grammaticalization*. Cambridge: Cambridge University Press.
Hellwig, Birgit. 2020. *A grammar of Qaqet*. Berlin, Boston: De Gruyter Mouton.
Helmbrecht, Johannes. 2020a. Form and function of personal names. Dimensions of the morphosyntactic diversity. In Nataliya Levkovych & Julia Nintemann (eds.), *Aspects of the grammar of names. Empirical case studies and theoretical topics*, 1–24. München: LINCOM.
Helmbrecht, Johannes. 2020b. On the morphosyntax of personal names in Hoocąk (Sioux). In Nataliya Levkovych & Julia Nintemann (eds.), *Aspects of the grammar of names. Empirical case studies and theoretical topics*, 147–166. München: LINCOM.
Helmbrecht, Johannes. 2022. Proper names with and without definite articles. Preliminary results. In Javier Caro Reina & Johannes Helmbrecht (eds.), *Proper names versus common nouns. Morphosyntactic contrasts in the languages of the world*, 121–154. Berlin, Boston: De Gruyter Mouton.
Helmbrecht, Johannes, Lukas Denk, Sarah Thanner & Ilenia Tonetti. 2018. Morphosyntactic coding of proper names and its implications for the Animacy Hierarchy. In Sonja Cristofaro & Fernando Zúñiga (eds.), *Typological Hierarchies in Synchrony and Diachrony*, 377–402. Amsterdam: John Benjamins.
Hennoste, Tiit & Karl Pajusalu. 2009. *Eesti murded ja kohanimed*. Tallinn: Eesti Keele Sihtasutus.

Hessky, Regina, Sarolta László, Csilla Bernáth, Bertalan Ilker & Erzsébet Knipf (eds.). 2005. *Pons Wörterbuch Schule und Studium 2: Ungarisch-Deutsch*. Stuttgart: Klett.

Hewitt, B. George. 1989. Abkhaz. In B. George Hewitt (ed.), *The indigenous languages of the Caucasus. Volume 2: The North West Caucasian languages*, 38–88. Delmar/NY: Caravan.

Himmelmann, Nikolaus P. 2005. The Austronesian languages of Asia and Madagascar. Typological characteristics. In Alexander Adelaar & Nikolaus P. Himmelmann (eds.), *The Austronesian languages of Asia and Madagascar*, 110–181. London, New York: Routledge.

Hockett, Charles. 1958. *A course in modern linguistics*. New York: Macmillan.

Hoekstra, Jarich. 2018. Frisian genitives. From Old Frisian to modern dialects. In Tanja Ackermann, Horst Simon & Christian Zimmer (eds.), *Germanic genitives*, 37–64. Amsterdam, Philadelphia: Benjamins.

Hough, Carole (ed.). 2016. *The Oxford handbook of names and naming*. Oxford: Oxford University Press.

Hualde, José Ignacio & Jon Ortiz de Urbina. 2003. *A Grammar of Basque*. Berlin, New York: Mouton de Gruyter.

Hyslop, Gwendolyn. 2017. *A grammar of Kurtöp*. Leiden, Boston: Brill.

Iordan, Iorgu & Vladimir Robu. 1978. *Limba română contemporană*. Bucureşti: Editura didactică şi pedagogică.

Jeffay, Elisheva & Susan Rothstein. 2022. On personal names in construct states in Modern and Biblical Hebrew. In Javier Caro Reina & Johannes Helmbrecht (eds.), *Proper names versus common nouns. Morphosyntactic contrasts in the languages of the world*, 155–174. Berlin, Boston: De Gruyter Mouton.

Jensen, John Thayer. 1977. *Yapese reference grammar*. Honolulu: University of Hawaii Press.

Kämpfe, Hans-Rainer & Alexander P. Volodin. 1995. *Abriß der tschuktschischen Grammatik auf der Basis der Schriftsprache*. Wiesbaden: Harrassowitz.

Karlsson, Fred. 1984. *Finnische Grammatik*. Hamburg: Buske.

Karnowski, Pawel & Jürgen Pafel. 2005. Wie anders sind Eigennamen? *Zeitschrift für Sprachwissenschaft* 24(1). 45–66.

Kempf, Luise, Damaris Nübling & Mirjam Schmuck. 2020. Warum eine Linguistik der Eigennamen? In Luise Kempf, Damaris Nübling & Mirjam Schmuck (eds.), *Linguistik der Eigennamen*, 1–18. Berlin, New York: De Gruyter.

Kimball, Geoffrey D. 1991. *Koasati grammar*. Lincoln, London: University of Nebraska Press.

Kolde, Gottfried. 1995. Grammatik der Eigennamen (Überblick). In Ernst Eichler, Gerold Hilty, Heinrich Löffler, Hugo Steger & Ladislav Zgusta (eds.), *Namenforschung*. 1. Halbband, 400–408. Berlin: De Gruyter.

Koptjevskaja-Tamm, Maria. 2002. Adnominal possession in the European languages. Form and function. *Sprachtypologie und Universalienforschung* 55(2). 141–172.

Kress, Bruno. 1982. *Isländische Grammatik*. Leipzig: Enzyklopädie.

Krishnamurti, Bh. & Brett A. Benham. 2020. Koṇḍa. In Sanford B. Steever (ed.), *The Dravidian languages*, 296–324. London, New York: Routledge.

Kukuczka, Elena. 1982. *Verwandtschaft, Körperteile und Besitz. Zur Possession im Tamil*. Köln: Institut für Sprachwissenschaft.

Kuteva, Tania, Bernd Heine, Bo Hong, Haiping Long, Heiko Narrog & Seongha Rhee. 2019. *World Lexicon of Grammaticalization*. Cambridge: Cambridge University Press.

Lafitte, Pierre. 1998. *Grammaire basque (navarro-labourdin littéraire)*. Donostia: Elkarlanean. [1st edition: 1944]

Langacker, Ronald W. 2008. *Cognitive grammar. A basic introduction*. Oxford: Oxford University Press.

Launey, Michel. 2003. *Awna parikwaki. Introduction à la langue palikur de Guyane et de l'Amapá*. Paris: IRD.

Launey, Michel. 2011. *An introduction to Classical Nahuatl.* 2011. Cambridge: Cambridge University Press.
Le Bihan, Hervé. 2015. The grammatical status and syntax of toponyms and anthroponyms. The case of Breton and Welsh. In Jonas Löfström & Betina Schnabel-Le Corre (eds.), *Challenges in synchronic toponymy. Structure, context, use*, 195–202. Tübingen: Narr Francke Attempto.
Lehmann, Christian. 1995. *Thoughts on grammaticalization.* München, Newcastle: Lincom Europa.
Lehmann, Christian. 1998. *Possession in Yucatec Maya. Structures, function, typology.* München, Newcastle: Lincom Europa.
Lestrade, Sander. 2010a. Finnish case alternating adpositions. A corpus study. *Linguistics* 48 (3), 603–628.
Lestrade, Sander. 2010b. *The Space of Case.* Nijmegen: Radboud Universiteit.
Levinson, Stephen C. 2003. *Space in language and cognition. Explorations in cognitive diversity.* Cambridge: Cambridge University Press.
Levkovych, Nataliya & Julia Nintemann. 2020. Preface. In Nataliya Levkovych & Julia Nintemann (eds.), *Aspects of the grammar of names. Empirical case studies and theoretical topics*, v–vii. München: LINCOM.
Lindow, Wolfgang, Dieter Möhn, Hermann Niebaum, Dieter Stellmacher, Hans Taubken & Jan Wirrer. 1998. *Niederdeutsche Grammatik.* Leer: Schuster.
Löbel, Elisabeth. 2002. The word class 'noun'. In D. Alan Cruse, Franz Hundsnurscher, Michael Job & Peter Rolf Lutzeier (eds.), *Lexikologie.* 1. Halbband, 588–597. Berlin: De Gruyter.
Lockwood, W.B. 1977. *An introduction to modern Faroese.* Tórshavn: Føroya Skúlabókagrunnur.
Loporcaro, Michele. 2018. *Gender from Latin to Romance. History, geography, typology.* Oxford: Oxford University Press.
Luraghi, Silvia. 2009. The evolution of local cases and their grammatical equivalent in Greek and Latin. In Jóhanna Barðdal & Shobhana L. Chelliah (eds.), *The Role of Semantic, Pragmatic, and Discourse Factors in the Development of Case*, 283–305. Amsterdam, Philadelphia: Benjamins.
Lutkat-Lõik, Florence-Silvia & Cornelius Hasselblatt. 2005. *Estnisch intensiv! Das Lehrbuch der estnischen Sprache.* Hamburg: Bibliotheca Baltica.
Lynch, John. 2011. Iaai. In John Lynch, Malcolm Ross & Terry Crowley (eds.), *The Oceanic languages*, 776–791. London, New York: Routledge.
Lynch, John, Malcolm Ross & Terry Crowley. 2011. Typological overview. In John Lynch, Malcolm Ross & Terry Crowley (eds.), *The Oceanic languages*, 34–53. London, New York: Routledge.
Malau, Catriona. 2016. *A grammar of Vurës, Vanuatu.* Boston, Berlin: De Gruyter Mouton.
Mallinson, Graham. 1986. *Rumanian.* Londo, Sydney: Croom Helm.
Matsumori, Akiko & Takuichiro Onishi. 2017. Japanese dialects. Focusing on Tsuruoka and Ei. In Nicolas Tranter (ed.), *The languages of Japan and Korea*, 313–348. London, New York: Routledge.
Matthews, Stephen & Virginia Yip. 2011. *Cantonese. A comprehensive grammar.* London, New York: Routledge.
Mauri, Caterian & Andrea Sansò. 2019. Nouns & Co. Converging evidence in the analysis of associative plurals. *Language Typology and Universals (STUF)* 72(4). 603–626.
Mazzitelli, Lidia Federica. 2015. *The expression of predicative possession. A comparative study of Belarusian and Lithuanian.* Berlin, Boston: De Gruyter Mouton.
Mbeje, Audrey N. 2005. *Zulu learners' reference grammar.* Madison: National African Languages Resource Center – University of Wisconsin.
McCarthy, John & Alan Prince. 1986. *Prosodic morphology.* New Brunswick: Rutgers University.
McCracken, Chelsea. 2021. *A grammar of Belep.* Berlin, Boston: De Gruyter Mouton.
Mikesy, Sándor. 1978. *Ungarisches Lehrbuch.* Leipzig: Enzyklopädie.

Mohtaschemi-Virkkunen, Mirja. 1970. *Finnische Texte*. Göppingen: Kümmerle.
Moravcsik, Edith A. 2013. *Introducing language typology*. Cambridge: Cambridge University Press.
Möhlig, Wilhelm J.G. & Bernd Heine. 1999. *Swahili Grundkurs*. Köln: Köppe.
Mojapelo, Mampaka L. 2009. Morphology and semantics of proper names in Northern Sotho. *South African Journal of African Languages* 29 (2). 185–194.
Nichols, Johanna. 1994. Chechen. In Rieks Smeets (ed.), *The indigenous languages of the Caucasus. Volume 4: North East Caucasian languages, part 2*, 1–78. Delmar/NY: Caravan.
Nintemann, Julia & Nicole Hober. In press. On the morphosyntax of place names vs. common nouns in pidgins and creoles: The encoding of two types of Ground in Goal and Source constructions. In Nataliya Levkovych (ed.), *Diversity in contact*. Berlin, Boston: De Gruyter.
Nintemann, Julia, Maja Robbers & Nicole Hober. 2020. *Here – hither – hence and related categories. A cross-linguistic study*. Berlin, Boston: De Gruyter Mouton.
Nübling, Damaris. 2005. Zwischen Syntagmatik und Paradigmatik. Grammatische Eigennamenmarker und ihre Typologie. *Zeitschrift für Germanistische Linguistik* 33. 25–56.
Nübling, Damaris. 2012. Auf dem Wege zu Nicht-Flektierbaren. Die Deflexion der deutschen Eigennamen diachron und synchron. In Björn Rothstein (ed.), *Nicht-flektierende Wortarten*, 224–246. Berlin, New York: De Gruyter.
Nübling, Damaris. 2017a. The growing distance between proper names and common nouns in German. On the development of onymic schema constancy. *Folia Linguistica* 51(2). 341–367.
Nübling, Damaris. 2017b. Funktionen neutraler Genuszuweisung bei Personennamen und Personenbezeichnungen im germanischen Vergleich. In Johannes Helmbrecht, Damaris Nübling & Barbara Schlücker (eds.), *Namengrammatik*, 173–211. Hamburg: Buske.
Nübling, Damaris, Fabian Fahlbusch & Rita Heuser. 2015. *Namen. Eine Einführung in die Onomastik*. 2nd edition. Tübingen: Narr.
Nyström, Staffan. 2016. Names and meaning. In Carole Hough (ed.), *The Oxford handbook of names and naming*, 39–51. Oxford: Oxford University Press.
Olsson, Bruno. 2022. *A grammar of Coastal Marind*. Berlin, Boston: De Gruyter Mouton.
Peţan, Aurora. 2001. Gen personal. In Marius Sala (ed.), *Enciclopedia limbii române*, 234–235. Bucureşti: Univers enciclopedic.
Petersen, Hjalmar P. & Renata Szczepaniak. 2018. The development of non-paradigmatic linking elements in Faroese and the decline of the genitive case. In Tanja Ackermann, Horst Simon & Christian Zimmer (eds.), *Germanic genitives*, 115–147. Amsterdam, Philadelphia: Benjamins.
Peterson, John. 2011. *A grammar of Kharia. A South Munda language*. Leiden, Boston: Brill.
Plungian, Vladimir. 1995. *Dogon*. München, Newcastle: Lincom Europa.
Polian, Gilles. 2017. Morphology. In Judith Aissen, Nora C. England & Roberto Zavala Maldonado (eds.), *The Mayan languages*, 201–225. London, New York: Routledge.
Popkema, Jan. 2018. *Grammatica fries*. Leeuwarden: Afûk.
Poulsen, Morten. 2002. *„Meinem Vater sein Haus" – Konstruktioner med resumptivt possessivpronomen i tysk og beslægtede sprog*. Aarhus: Aarhus Universitet, Institut for Germansk Filologi.
Rasoloson, Janie & Carl Rubino. 2011. Malagasy. In Alexander Adelaar & Nikolaus P. Himmelmann (eds.), *The Austronesian languages of Asia and Madagascar*, 456–488. London, New York: Routledge.
Renck, G.L. 1975. *A grammar of Yagaria*. Canberra: The Australian National University.
Rijkhoff, Jan. 2002. *The noun phrase*. Oxford: Oxford University Press.
Roberts, John R. 1987. *Amele*. London, New York, Sydney: Croom Helm.
Rodrigues Aristar, Anthony. 1997. Marking and hierarchy types and the grammaticalization of case-markers. *Studies in Language* 21(2). 313–368.

Rodríguez Guaján, José Obispo. 1994. *Rutz'ib'axik ri Kaqchikel. Manual de redacción Kaqchikel.* Guatemala: Cholsamaj.
Ross, Malcolm. 2011a. Mussau. In John Lynch, Malcolm Ross & Terry Crowley (eds.), *The Oceanic languages*, 148–164. London, New York: Routledge.
Ross, Malcolm. 2011b. Kaulong. In John Lynch, Malcolm Ross & Terry Crowley (eds.), *The Oceanic languages*, 387–409. London, New York: Routledge.
Rubino, Carl Ralph Galvez. 2000. *Ilocano dictionary and grammar. Ilocano-English, English-Ilocano.* Honolulu: University of Hawaii.
Rubino, Carl. 2011. Iloko. In Alexander Adelaar & Nikolaus P. Himmelmann (eds.), *The Austronesian languages of Asia and Madagascar*, 326–349. London, New York: Routledge.
Saeed, John. 1999. *Somali.* Amsterdam, Philadelphia: Benjamins.
Salaberri, Iker. 2022. D-marking on Basque personal names from a synchronic and diachronic perspective. In Javier Caro Reina & Johannes Helmbrecht (eds.), *Proper names versus common nouns. Morphosyntactic contrasts in the languages of the world*, 205–236. Berlin, Boston: De Gruyter Mouton.
Sarvasy, Hannah S. 2017. *A grammar of Nungon. A Papuan language of Northeast New Guinea.* Leiden, Boston: Brill.
Schlücker, Barbara & Tanja Ackermann. 2017. The morphosyntax of proper names. An overview. *Folia Linguistica* 51(2). 483–504.
Schlücker, Barbara, Damaris Nübling & Johannes Helmbrecht. 2017. Einleitung. In Johannes Helmbrecht, Damaris Nübling & Barbara Schlücker (eds.), *Namengrammatik*, 5–10. Hamburg: Buske.
Schuster, Susanne. 2020. *HABEN ODER NICHT HABEN. DIACHRONE BESCHREIBUNG UND ANALYSE DES ISLÄNDISCHEN POSSESSIONSSYSTEMS.* Heidelberg: Winter.
Seiler, Hansjakob. 1983a. Namengebung als eine Technik zur sprachlichen Erfassung von Gegenständen. In Manfred Faust, Roland Harweg, Werner Wienold & Götz Lehfeldt (eds.), *Allgemeine Sprachwissenschaft, Sprachtypologie und Textlinguistik*, 149–156. Tübingen: Narr.
Seiler, Hansjakob. 1983b. *Possession as an operational dimension of language.* Tübingen: Narr.
Shivutse, Ernesti. 1972. *Suaheli für Sie.* München: Hueber.
Smailus, Ortwin. 1989a. *Vocabulario en lengua castellana y guatemalteca que se llama Cakchiquel chi. Tomo I, Parte 1: Grámatica.* Hamburg: Wayasbah.
Smailus, Ortwin. 1989b. *Vocabulario en lengua castellana y guatemalteca que se llama Cakchiquel chi. Tomo II, Parte 2a: Lexicología A-J.* Hamburg: Wayasbah.
Smailus, Ortwin. 1989c. *Vocabulario en lengua castellana y guatemalteca que se llama Cakchiquel chi. Tomo III, Parte 2b: Lexicología K-Z.* Hamburg: Wayasbah.
Song, Jae Jung. 2018. *Linguistic typology.* Oxford: Oxford University Press.
Sonnenhauser, Barbara & Patrizia Noel Aziz Hanna (eds.). 2013. *Vocative! Addressing between System and Performance.* Berlin, Boston: De Gruyter Mouton.
Spencer, Andrew. 2007. Realization-based morphosyntax. The German genitive. In Patrick O. Steinkrüger & Manfred Krifka (eds.), *On inflection*, 173–218. Berlin, New York: Mouton de Gruyter.
Staksberg, Marius. 1996. Sa-possessiv. *Málting* 18(3). 28–34.
Stan, Camelia, Gabriela Pană Dindelegan, Alexandru Nicolae, Raluca Brăescu, Andra Vasilescu. 2016. The nominal phrase. In Gabriela Pană Dindelegan (ed.), *The syntax of Old Romanian*, 288–393. Oxford: Oxford University Press.
Stassen, Leon. 2009. *Predicative possession.* Oxford: Oxford University Press.
Stoebke, Renate. 1968. *Die Verhältniswörter in den ostseefinnischen Sprachen.* Bloomington: Indiana University.

Stolz, Thomas. 2004. Possessions in the Far North. A glimpse of the alienability correlation in modern Icelandic. In Premper Waldfried (ed.), *Dimensionen und Kontinua. Beiträge zu Hansjakob Seilers Universalienforschung*, 73–96. (= Diversitas Linguarum 4). Bochum: Universitätsverlag Dr. N. Brockmeyer.

Stolz, Thomas. 2012. Europäische Besitzungen. Zur gespaltenen Possession im europäischen Sprachvergleich. In Lutz Gunkel & Gisela Zifonun (eds.), *Deutsch im Sprachvergleich. Grammatische Kontraste und Konvergenzen*, 41–73. Berlin: De Gruyter.

Stolz, Thomas. 2019. Differentielle Namenkörperschonung. Zur Anlautmutation von Personen- und Ortsnamen im heutigen Walisischen (mit einem Ausblick auf seine keltischen Verwandten). *Beiträge zur Namenforschung* 54(1). 15–70.

Stolz, Thomas. 2020. Is there anything wrong with *iya*? On morphosyntactic issues connected to place names in Chamorro. In Nataliya Levkovych & Julia Nintemann (eds.), *Aspects of the grammar of names. Empirical case studies and theoretical topics*, 53–145. München: LINCOM.

Stolz, Thomas & Sabine Gorsemann. 2001. Pronominal possession in Faroese and the parameters of alienability/inalienability. *Studies in Language* 25. 557–599.

Stolz, Thomas, Sonja Kettler, Cornelia Stroh & Aina Urdze. 2008. *Split possession. An areal-linguistic study of the alienability correlation and related phenomena in the languages of Europe*. Amsterdam, Philadelphia: John Benjamins.

Stolz, Thomas, Sander Lestrade & Christel Stolz. 2014. *The Crosslinguistics of Zero-Marking of Spatial Relations*. Berlin, Boston: De Gruyter Mouton.

Stolz, Thomas & Nataliya Levkovych. 2019a. Toponomastics meets linguistic typology. Glimpses of Special Toponymic Grammar from Aromanian and sundry languages. *Onomastica Uralica* 11. 43–61.

Stolz, Thomas & Nataliya Levkovych. 2019b. Absence of material exponence. *Language Typology and Universals – STUF* 72(3). 373–400.

Stolz, Thomas & Nataliya Levkovych. 2020a. Zwischen Ortsnamenbildung und Relationsmarkierung. Strukturelle Ambiguitäten, Grauzonen und Übergänge. *Beiträge zur Namenforschung* 55(1). 1–25.

Stolz, Thomas & Nataliya Levkovych. 2020b. Grammatical versus onymic classifiers. First thoughts about a potentially interesting topic. In Nataliya Levkovych & Julia Nintemann (eds.), *Aspects of the grammar of names. Empirical case studies and theoretical topics*, 167–180. München: LINCOM.

Stolz, Thomas & Nataliya Levkovych. 2022. On *Special Onymic Grammar (SOG)*. Definiteness markers in Fijian and selected Austronesian languages. In Javier Caro Reina & Johannes Helmbrecht (eds.), *Proper names versus common nouns. Morphosyntactic contrasts in the languages of the world*, 237–263. Berlin, Boston: De Gruyter Mouton.

Stolz, Thomas, Nataliya Levkovych & Aina Urdze. 2017a. Die Grammatik der Toponyme als typologisches Forschungsfeld. Eine Pilotstudie. In Johannes Helmbrecht, Damaris Nübling & Barbara Schlücker (eds.), *Namengrammatik*, 121–146. Hamburg: Buske.

Stolz, Thomas, Nataliya Levkovych & Aina Urdze. 2017b. When zero is just enough. . . In support of a Special Toponymic Grammar in Maltese. *Folia Linguistica* 51(2). 453–482.

Stolz, Thomas, Nataliya Levkovych & Aina Urdze. 2018. La morfosintassi dei toponimi in prospettiva tipologica. In Giuseppe Brincat & Sandro Caruana (eds.), *Tipologia e 'dintorni'. Il metodo tipologico alla intersezione di piani d'analisi*, 307–324. Roma: Bulzoni.

Suttles, Wayne. 2004. *Musqueam reference grammar*. Vancouver, Toronto: University of British Columbia Press.

Svorou, Soteria. 1993. *The grammar of space*. Amsterdam, Philadelphia: Benjamins.

Szczepaniak, Renata. 2005. Onymische Suffixe als Signal der Proprialität – das Polnische als Paradebeispiel. In Eva Brylla & Mats Wahlberg (eds.), *Proceedings of the 21st International Conference of Onomastic Sciences (ICOS)*, 295–308. Uppsala: Språk- och folkminnesinstitutet.

Taavitsainen, Irma & Andreas H. Jucker. 2016. Forms of address. In Carole Hough (ed.), *The Oxford handbook of names and naming*, 427–437. Oxford: Oxford University Press.
Talmy, Leonard. 1985. Lexicalization patterns. Semantic structure in lexical forms. In Timothy Shopen (ed.), *Language Typology and Syntactic Description 3: Grammatical Categories and the Lexicon*, 57–149. Cambridge: Cambridge University Press.
Tauli, Valter. 1983. *Standard Estonian Grammar. Part II: Syntax*. Uppsala: Acta Universitatis Upsaliensis.
Taylor, Charles. 1985. *Nkore-Kiga*. London, Sydney, Dover/NH: Croom Helm.
Taylor, John R. 1989. *Linguistic categorization. Prototypes in linguistic theory*. Oxford: Clarendon.
Thanner, Sarah. 2019. *Personennamen und die Belebtheitshierarchie. Zur morphosyntaktischen Kodierung von Personennamen in Split-Ergativsprachen*. Regensburg: Universität Regensburg, Lehrstuhl für Allgemeine und Vergleichende Sprachwissenschaft.
Thornton, Anna M. 1996. On some phenomena of prosodic morphology in Italian: accorciamenti, hypocoristics and prosodic delimitation. *Probus* 8(1). 81–112.
Thornton, Anna M. 2012. Overabundance in Italian verb morphology and its interaction with other non-canonical phenomena. In Thomas Stolz, Hitomi Otsuka, Aina Urdze & Johan van der Auwera (eds.), *Irregularity in Morphology (and beyond)*, 251–270. Berlin: Akademie.
Thráinsson, Höskuldur. 2010. *The syntax of Icelandic*. Cambridge: Cambridge University Press.
Thráinsson, Höskuldur, Hjalmar P. Petersen, Jógvan í Lon Jacobsen & Zakaris Svabo Hansen. 2004. *Faroese. An overview and reference grammar*. Tórshavn: Føroya Fróðskaparfelag.
Tompa, József. 1972. *Kleine ungarische Grammatik*. Leipzig: Enzyklopädie.
Touratier, Christian. 2013. *Lateinische Grammatik. Linguistische Einführung in die lateinische Sprache*. Darmstadt: Wissenschaftliche Buchgesellschaft.
Treis, Yvonne. 2008. *A grammar of Kambaata. Part 1: Phonology, nominal morphology, non-verbal predication*. Köln: Köppe.
Tsukida, Naomi. 2011. Seediq. In Alexander Adelaar & Nikolaus P. Himmelmann (eds.), *The Austronesian languages of Asia and Madagascar*, 291–325. London, New York: Routledge.
Van de Velde, Mark & Odette Ambouroue. 2011. The grammar of Orungu proper names. *Journal of African Languages and Linguistics* 32(1). 113–141.
Van den Toorn, M.C. 1977. *Nederlandse grammatica*. Groningen: Wolters-Nordhoff.
Van Langendonck, Willy. 2007. *Theory and typology of proper names*. Berlin, New York: Mouton de Gruyter.
Van Langendonck, Willy & Mark van de Velde. 2016. Names and grammar. In Carole Hough (ed.), *The Oxford handbook of names and naming*, 17–38. Oxford: Oxford University Press.
Velupillai, Viveka. 2012. *An introduction to linguistic typology*. Amsterdam, Philadelphia: Benjamins.
Viitso, Tiit-Rein. 2003. Structure of the Estonian language. Phonology, morphology and word formation. In Mati Erelt (ed.), *Estonian Language*, 9–129. Tallinn: Estonian Academic Publishers.
Villasante, Fr. Luís. 1972. *La declinación del vasco literario común*. Oñate: Editorial Franciscana Aranzazu.
Whaley, Lindsay J. 1997. *Introduction to typology. The unity and diversity of language*. London, New Dehli: Sage.
Zipf, George K. 1968. *The psycho-biology of language. An introduction to dynamic philology*. Cambridge/Mass.: MIT Press [1st edition 1935].

Appendix I: List of languages (n = 85)

Language	Affiliation	Continent
Abkhaz	Abkhaz-Adyge	Europe
Alaskan Yup'ik	Eskimo-Aleut	Americas
Amele	Nuclear Trans New Guinea, Madang	Oceania
Ancient Greek	Indo-European, Greek	Europe
Arosi	Austronesian, Oceanic	Oceania
Bali-Vitu	Austronesian, Oceanic	Oceania
Basque	Isolate	Europe
Belep	Austronesian, Oceanic	Oceania
Buol	Austronesian, Greater Central Philippine	Asia
Cantonese	Sino-Tibetan, Sinitic	Asia
Catalan	Indo-European, Romance	Europe
Cebuano	Austronesian, Greater Central Philippine	Asia
Cèmuhî	Austronesian, Oceanic	Oceania
Chechen	Nakh-Daghestanian, Nakh	Europe
Chukchi	Chukotko-Kamchatkan	Asia
Coastal Marind	Anim	Asia
Dogon	Dogon	Africa
Dutch	Indo-European, Germanic	Europe
Eastern Armenian	Indo-European, Armenic	Europe
Estonian	Uralic, Finnic	Europe
Faroese	Indo-European, Germanic	Europe
Finnish	Uralic, Finnic	Europe
Gela	Austronesian, Oceanic	Oceania
German	Indo-European, Germanic	Europe
Halkomelem	Salishan	Americas
Herero	Atlantic-Congo, Bantu	Africa
Hungarian	Uralic, Hungaric	Europe
Iaai	Austronesian, Oceanic	Oceania
Icelandic	Indo-European, Germanic	Europe
Iloko	Austronesian, Northern Luzon	Asia
Italian	Indo-European, Romance	Europe
Japanese (Ei)	Japonic	Asia
Japanese (Tsuruoka)	Japonic	Asia
Kambaata	Afro-Asiatic, Cushitic	Africa
Kambera	Austronesian, Bima-Lembata	Asia
Kaqchikel	Mayan	Americas
Kaulong	Austronesian, Oceanic	Oceania
Kera	Afro-Asiatic, Chadic	Africa
Kharia	Austroasiatic, Mundaic	Asia
Kimaragang	Austronesian, North Borneo Malayo-Polynesian	Asia
Koasati	Muskogean	Americas
Kobon	Nuclear Trans New Guinea, Madang	Oceania

(continued)

Language	Affiliation	Continent
Koṇḍa	Dravidian	Asia
Kryts	Nakh-Daghestanian, Lezgic	Europe
Lango	Nilotic	Africa
Lao	Tai-Kadai, Daic	Asia
Latin	Indo-European, Latinic	Europe
Lezgian	Nakh-Daghestanian, Lezgic	Europe
Low German	Indo-European, Germanic	Europe
Malagasy	Austronesian, Basap-Greater Barito	Africa
Manx	Indo-European, Celtic	Europe
Maori	Austronesian, Oceanic	Oceania
Marquesan	Austronesian, Oceanic	Oceania
Mauwake	Nuclear Trans New Guinea, Madang	Oceania
Mian	Nuclear Trans New Guinea, Ok	Oceania
Mori Bawah	Austronesian, Celebic	Asia
Mussau	Austronesian, Oceanic	Oceania
Nadrogā	Austronesian, Oceanic	Oceania
Nivkh	Isolate	Asia
Nkore-Kiga	Atlantic-Congo, Bantu	Africa
Norwegian	Indo-European, Germanic	Europe
Nungon	Nuclear Trans New Guinea, Finisterre-Huon	Oceania
Palikur	Arawakan	Americas
Qaqet	Baining	Oceania
Romanian	Indo-European, Romance	Europe
Rotuman	Austronesian, Oceanic	Oceania
Roviana	Austronesian, Oceanic	Oceania
Sanzhi Dargwa	Nakh-Daghestanian, Dargwic	Europe
Seediq	Austronesian, Atayalic	Asia
Siar	Austronesian, Oceanic	Oceania
Somali	Afro-Asiatic, Cushitic	Africa
Southern Peruvian Quechua	Quechuan	Americas
Spanish	Indo-European, Romance	Europe
Swahili	Atlantic-Congo, Bantu	Africa
Tagalog	Austronesian, Greater Central Philippine	Asia
Taiof	Austronesian, Oceanic	Oceania
Tamambo	Austronesian, Oceanic	Oceania
Tariana	Arawakan	Americas
Tswana	Atlantic-Congo, Bantu	Africa
Tucano	Tucanoan	Americas
Vurës	Austronesian, Oceanic	Oceania
West Frisian	Indo-European, Germanic	Europe
Yagaria	Nuclear Trans New Guinea, Goroka	Oceania
Yapese	Austronesian, Oceanic	Asia
Zulu	Atlantic-Congo, Bantu	Africa

Appendix II: Map

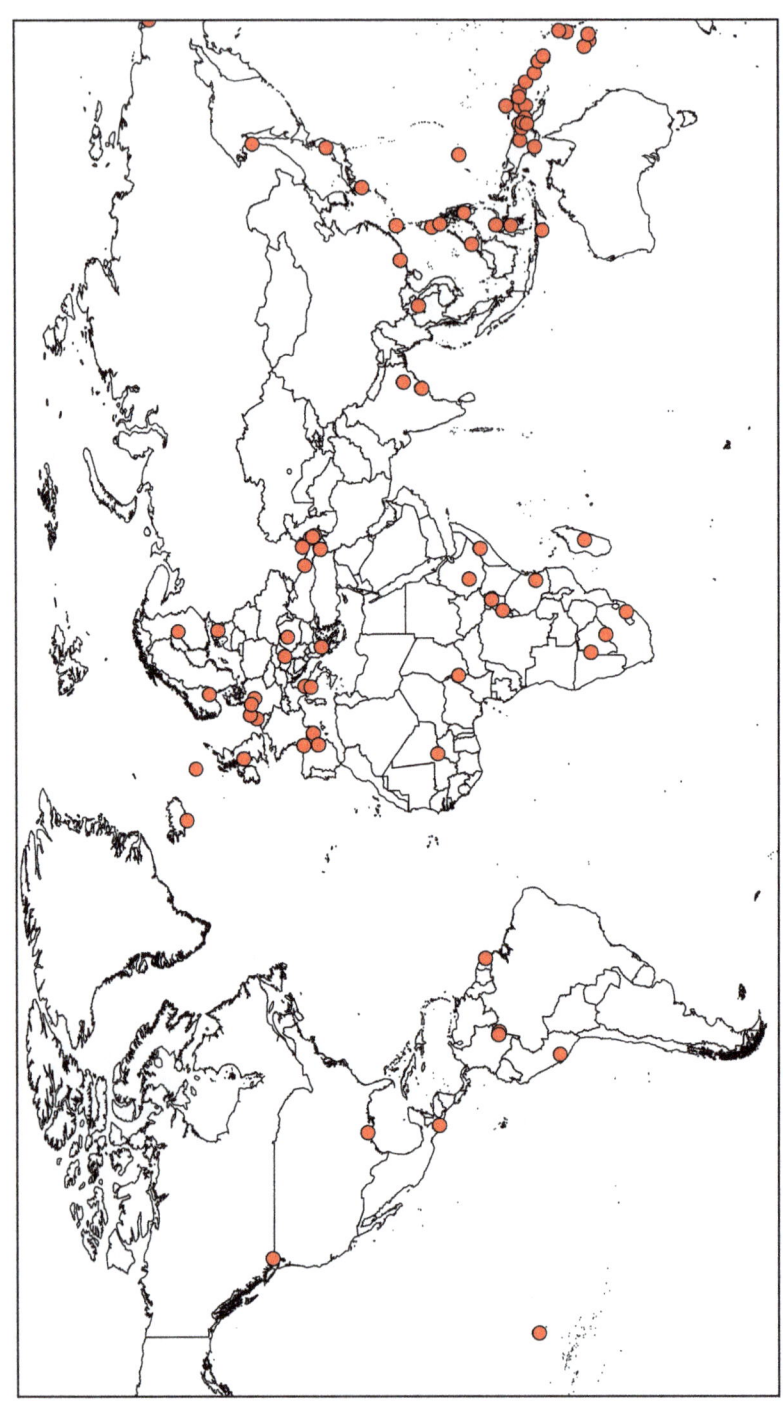

Index of authors

Ackermann, Tanja 22, 65
Adelaar, Willem F.H. 45
Aikhenvald, Alexandra Y. 3, 15, 31, 44, 45, 100, 194
Ainiala, Terhi 206
Aldrin, Emilia 206
Allerton, D.J. 206
Ambouroue, Odette 37
Anderson, John M. 19, 20, 22, 24–26, 30, 36, 37, 44, 103, 119, 151, 170, 175
Authier, Gilles 62

Baerman, Matthew 36, 109
Bánhidi, Zoltán 119, 121
Bárczi, Géza 115
Bastin, Yvonne 224
Bauer, Gerhard 23
Bauer, Winifred 47, 48
Becker, Laura 15, 55, 103, 202, 215
Bendel, Christiane 149, 150
Benham, Brett A. 45
Berchtold, Simone 37
Berghäll, Liisa 48
Beyrer, Arthur 92, 94–97
Blake, Barry J. 44
Boisserie, Étienne 120
Booij, Geert 20, 67, 70
Bornemann, Eduard 160, 161, 163–165
Bosque, Ignacio 205
Boyeldieu, Pascal 38
Brauner, Siegmund 183–192
Braunmüller, Kurt 70, 78
Broderick, George 60
Brown, Dunstan 20
Brown, Penelope 206
Brown, R. McKenna 203

Caragiu Marioțeanu, Matilda 99
Caro Reina, Javier 22, 37, 38, 93, 94, 200, 201
Chantraine, Pierre 161
Chappell, Hilary 29
Chumakina, Marina 20
Cojocaru, Dana 97, 98
Comrie, Bernard 106, 207

Corbett, Greville G. 15, 20, 22, 30, 36, 92, 93, 109, 156, 219, 225
Cornilescu, Alexandra 99
Corston-Oliver, Simon 211–213
Creissels, Denis 25
Croft, William 25, 26, 42, 43, 102, 207, 208, 211, 213
Crowley, Terry 214
Cysouw, Michael 19, 104

D'hulst, Yves 38
Dahl, Östen 29
Dammel, Antje 22, 37
Daniel, Michael 31, 44, 45
Davies, Jon 59
De Rijk, Rudolf P.G. 149, 150, 152, 153, 155, 156
Delsing, Lars-Olof 82
Dimitrescu, Florica 99
Dixon, R.M.W. 1, 30
Doleschal, Ursula 226
Draskau, Jennifer Kewley 60

Ebert, Karen H. 54
Eichler, Ernst 22, 23
Eisenbeiß, Sonja 66
Elderkin, Edward D. 229
Enfield, N.J. 45
Erelt, Tiiu 135, 144, 145, 147
Espiritu, Precy 2–13

Fedden, Sebastian 55
Flores, P. F. Ildefonso Joseph 205
Forker, Diana 41, 42
Fried, Mirjam 100

Geerts, G. 217
Gemoll, Wilhelm 165, 169
Gildersleeve, B.L. 170
Givón, T. 207
Goldap, Christel 197
Gönczöl, Ramona 97, 98
Good, Jeff 19
Gorsemann, Sabine 75, 77, 79

Gowlett, Derek 225
Guzmán Naranjo, Matías 103

Handschuh, Corinna 22, 30, 31, 43, 44
Harries, L. 184
Harweg, Roland 29
Haspelmath, Martin 18, 20, 40, 103, 106–110, 227, 228
Hasselblatt, Cornelius 140
Heidenkummer, Alexandra 38
Heine, Bernd 9, 100, 101, 182, 184, 187, 190, 192
Hellwig, Birgit 47, 56, 57
Helmbrecht, Johannes 22, 38, 39, 43, 102
Hennoste, Tiit 136
Herms, Irmtraud 183–192
Hessky, Regina 115
Hewitt, B. George 51
Himmelmann, Nikolaus P. 17
Hober, Nicole 38
Hockett, Charles 32, 92
Hoekstra, Jarich 69
Hough, Carole 18, 23
Hualde, José Ignacio 149, 150, 152, 154, 155, 157, 158
Hyslop, Gwendolyn 18

Iordan, Iorgu 94, 96

Jeffay, Elisheva 38
Jensen, John Thayer 44
Jucker, Andreas H. 206

Kämpfe, Hans-Rainer 52
Karlsson, Fred 125–129, 131
Karnowski, Pawel 42
Kimball, Geoffrey D. 54, 55
Kolde, Gottfried 23
Koptjevskaja-Tamm, Maria 29, 59
Kress, Bruno 86
Krishnamurti, Bh. 45
Kukuczka, Elena 29
Kuteva, Tania 201

Lafitte, Pierre 149, 155
Langacker, Ronald W. 197, 198
Launey, Michel 29, 39, 40
Le Bihan, Hervé 38

Lehmann, Christian 100, 204
Lestrade, Sander 125, 181, 208
Levinson, Stephen C. 197, 206
Levkovych, Nataliya 22, 25, 38, 48, 198, 210, 225
Lindow, Wolfgang 70
Löbel, Elisabeth 32
Lockwood, W.B. 76–79, 81
Lodge, Gonzalez 170
Loporcaro, Michele 92
Luraghi, Silvia 159–164, 172, 178, 179
Lutkat-Lõik, Florence-Silvia 140
Lynch, John 17, 211

Malau, Catriona 32, 34–36
Mallinson, Graham 94
Matsumori, Akiko 65
Matthews, Stephen 51, 52
Mauri, Caterian 31
Mazzitelli, Lidia Federica 100
Mbeje, Audrey N. 218
McCarthy, John 45
McCracken, Chelsea 51
McGregor, William 29
Mikesy, Sándor 114, 115, 120, 123
Möhlig, Wilhelm J.G. 182, 184, 187, 190, 192
Mohtaschemi-Virkkunen, Mirja 126, 127
Mojapelo, Mampaka L. 37
Moravcsik, Edith A. 31, 42

Nichols, Johanna 52
Nicolae, Alexandru 99
Nintemann, Julia 20, 22, 38
Noel Aziz Hanna, Patrizia 44
Nowak, Jessica 38
Nübling, Damaris 1, 4, 18–20, 26, 27, 29, 30, 32, 36, 37, 51, 65, 66, 102, 207, 224, 226, 229
Nyström, Staffan 197

Olsson, Bruno 52
Onishi, Takuichiro 65
Ortiz de Urbina, Jon 149, 150, 152, 154, 155, 157, 158

Pafel, Jürgen 42
Kempf, Luise 22, 42
Pajusalu, Karl 136
Peţan, Aurora 97

Petersen, Hjalmar P. 75
Peterson, John 54
Plungian, Vladimir 53
Polian, Gilles 205
Popkema, Jan 69
Poulsen, Morten 70
Prince, Alan 45

Rasoloson, Janie 62–64
Renck, G.L. 221, 222
Rijkhoff, Jan 42
Risch, Ernst 160, 161, 163–165
Roberts, John R. 61
Robu, Vladimir 94, 96
Rodrigues Aristar, Anthony 207
Rodríguez Guaján, José Obispo 203, 206
Ross, Malcolm 208, 209, 219, 220
Rothstein, Susan 38
Rubino, Carl Ralph Galvez 3, 6, 7, 13, 14, 62–64

Saeed, John 57, 58
Salaberri, Iker 38
Sansò, Andrea 31
Sarvasy, Hannah S. 28–31
Schlücker, Barbara 22
Schuster, Susanne 20, 82, 83, 100
Seiler, Hansjakob 32, 100
Shivutse, Ernesti 184, 191–193
Sims, Andrea D. 109
Smailus, Ortwin 204, 205
Smith, Norval 106
Song, Jae Jung 42
Sonnenhauser, Barbara 44
Spencer, Andrew 44, 45, 66
Staksberg, Marius 78, 82
Stan, Camelia 99, 150
Stassen, Leon 100

Stoebke, Renate 128
Stolz, Thomas 12, 25, 36–38, 46, 48, 59, 67, 71–73, 75, 77–79, 82, 83, 92, 106, 198, 199, 208, 210, 225, 229
Suttles, Wayne 215
Svorou, Soteria 197, 199
Szczepaniak, Renata 37, 75

Taavitsainen, Irma 206
Talmy, Leonard 5, 107, 197
Tauli, Valter 135, 137
Taylor, Charles 55, 56
Taylor, John R. 100
Thanner, Sarah 43, 102
Thornton, Anna M. 45, 118
Thráinsson, Höskuldur 75–84, 87, 91
Tompa, József 111–113, 115, 119, 121
Touratier, Christian 172, 177, 178
Treis, Yvonne 53, 54
Tsukida, Naomi 57

Van de Velde, Mark 1, 14, 23, 24, 34, 36, 37
Van den Toorn, M.C. 217
Van Langendonck, Willy 1, 14, 18–20, 22–26, 30, 34, 36, 37, 102
Velupillai, Viveka 42
Viitso, Tiit-Rein 133–135
Villasante, Fr. Luís 149, 150, 153, 154, 158
Volodin, Alexander P. 52

Whaley, Lindsay J. 42
Wohlgemuth, Jan 104

Yip, Virginia 51, 52

Zipf, George K. 229

Index of languages

Abkhaz 51, 58
Alaskan Yup'ik 45
Albanian 25
Amele 61, 103, 227
Aromanian 38
Arosi 17

Bali-Vitu 17
Basque 25, 38, 107, 149–153, 157–159, 182, 195, 196, 198, 225, 228
Belep 51
Breton 38
Buol 17

Cantonese 51, 52, 58
Catalan 200–206
– Balearic Catalan 37
Cebuano 17
Cèmuhî 17
Chamorro 12, 38, 39
Chechen 52, 58, 225
Chukchi 52, 58, 225
Classical Nahuatl 29, 36
Coastal Marind 52, 53, 58

Dogon 53, 58
Dutch 23, 26, 30, 67, 68, 70, 217, 218

Eastern Armenian 207
English 2, 3, 23, 24, 26, 30, 84, 112, 133, 150, 187, 188, 201
Estonian 111, 130, 135–149, 196, 225

Faroese 75–79, 81–83, 104, 225
Fijian 32, 38, 39
Finnish 111, 125–133, 135, 137, 139, 141, 148, 149, 182, 196
Flemish 26
French 24, 26, 37, 192, 229

Gela 17, 214, 215
German 2, 27, 36, 37, 65–68, 169, 183, 186, 192, 207, 219, 225, 229

Greek 24
– Ancient Greek 107, 159–165, 169–172, 181, 182, 195, 196, 225, 227, 228

Halkomelem 215, 216, 228
Hebrew 38, 39
Herero 229
Hoocąk 38
Hua 32
Hungarian 111–125, 128, 130–133, 135, 137, 139, 148, 149, 182, 195, 196, 225

Iaai 210, 211
Icelandic 76, 82–85, 87–92, 104
Iloko 2–4, 6–18, 31, 32, 34–36, 40, 42, 209, 225
Italian 45

Japanese 65

Kambaata 53, 54, 58
Kambera 17
Kaqchikel 200, 202–206
Kaulong 17, 208–210
Kera 54, 58
Kharia 54, 58
Kimaragang 17
Kirundi 23
Koasati 54, 55, 58, 225
Kobon 59
Koṇḍa 45
Kryts 62, 103, 225
Kurtöp 18

Lango 229
Lao 45
Latin 107, 159, 170–182, 195, 196, 201, 225
Lezgian 107–113, 115, 124, 149, 159, 182, 195, 196, 198, 225, 229
Low German 69–74, 104

Macedonian 25
Malagasy 17, 62–64
Maltese 38, 39

https://doi.org/10.1515/9783111331874-013

Manx 60, 61, 64
Maori 47, 48
Marquesan 17
Mauwake 48
Mian 55, 58
Mori Bawah 17
Mussau 219–223, 228

Nadrogã 17
Namia 23
Nivkh 45
Nkore-Kiga 55, 58
North Sotho 37
Norwegian 82
Nungon 28–31, 35, 36, 42, 225

Orungu 37

Palikur 39–42, 225
Polish 26, 37

Qaqet 47, 56–58

Romanian 65, 92–99, 104, 120, 225
Rotuman 17
Roviana 17, 211–213

Sanzhi Dargwa 41
Seediq 57, 58

Siar 17
Sinyar 38
Somali 57, 58
Southern Peruvian Quechua 45
Spanish 38, 203–206
Swahili 107, 182–187, 189–191, 193–196, 198, 218

Tagalog 17, 32
Taiof 17
Tamambo 17
Tariana 45
Tswana 225, 228
Tucano 45

Ughele 15

Vurës 32–36, 42, 225, 228

Welsh 38
West Frisian 69, 70

Yagaria 221–223
Yapese 44

Zone J Bantu 224, 225
Zulu 218, 219, 221, 225

Index of subjects

ablative 39–41, 52, 111–115, 125, 127, 133, 137, 140–142, 144–147, 150–156, 163, 171–175, 177–181, 188
absolutive 62, 150–153, 211–213
adessive 110–113, 125–129, 133, 137–139, 141, 143, 145, 180
adjective/adjectival 11, 30, 33, 48, 61, 72, 93, 164, 178
accusative 53, 75–79, 81, 83, 93, 94, 159–164, 166–168, 170–172, 174, 177, 178, 180, 196, 204, 228
adposition/adpositional 125, 196
adverb/adverbial 7, 39, 40, 79, 112, 127, 134, 135, 161, 163, 165, 166, 169, 182, 183, 186, 196, 207, 209
affix/affixal 52, 99, 153–157, 185, 198, 225
- affixation 154, 157, 185
allative 39, 41, 52, 111–113, 125–127, 129, 133, 137–139, 141, 143, 145–147, 150–156, 159, 161–163, 178, 180
animacy 29, 4, 70, 101, 102, 151, 155, 156, 158, 191–193
- animacy hierarchy 43, 102, 103
- animate 57, 72, 73, 77, 93, 101–103, 151–156, 158, 159, 180, 191, 202, 228
- inanimacy 150, 151
- inanimate 71–73, 101–103, 151–159, 179, 180, 184–188, 190–192, 213, 228
article 3–5, 8, 10–12, 14–16, 33, 35, 38, 48, 51, 55, 56, 62–64, 83, 95, 99, 104, 112, 125, 133, 188, 201, 202, 205, 211–215
associative plural/dual 16, 31, 51, 53, 55, 64, 78, 123, 132
asymmetry/asymmetric 163, 193, 207, 218, 227, 228

benefactive 138, 150, 151

canon/canonical 26, 32, 36, 81, 219, 224
- Canonical Morphology/Typology 15, 20, 58, 200, 224, 226, 227
- canonicity 22, 225
- noncanonicity 22, 54, 58, 224, 226

clitic/cliticisation 33, 35, 36, 53, 55, 64, 78, 79–81, 99, 164–170, 195, 220, 227
- enclitic 62, 63, 81, 99
- proclitic 62, 63, 96, 99, 218, 219
comitative 150, 151, 220
complement 7, 36, 53, 54, 66, 79, 80, 116–118, 128, 129, 137, 163, 166, 169, 171, 174, 175, 183, 211, 217, 220, 221
constraint 41, 60

dative 75, 76, 81, 95–97, 150, 151, 158–161, 163, 165, 166, 171, 172, 220
definite/definiteness 3, 14–16, 24, 25, 31, 37–39, 43, 51, 60, 63, 64, 76, 80, 93, 94, 99, 101, 104, 112, 125, 133, 150–158, 188, 200–202, 205, 208, 213
- indefinite/indefiniteness 3, 14, 16, 61, 63, 73, 94, 104, 125, 133, 150, 151, 153–155, 157, 158, 208, 213
deictic/deixis 13, 14, 16, 40, 41, 209, 215, 227
delative 111, 112, 115, 116, 122
declension 52, 66, 79–81, 119, 125, 128, 150, 154, 171–175, 177–180
demonstrative 30, 33, 53
deponency 109, 225
derivation/derivational 19, 35, 38, 54, 58, 106
diachronic/diachrony 12, 14, 20, 27, 28, 35, 37, 38, 65, 83, 99, 122, 147, 159, 179, 182, 200, 207, 2013
dialect/dialectal 45, 65, 154, 190, 200, 227
directive 150–156, 164
directional/directionality 41, 75, 164, 190, 192–194, 209, 210
dual 30, 31
dynamic 127, 142, 158, 163, 164, 167, 168, 171, 179, 180, 187, 190–193

elative 111–113, 115, 118, 121, 122, 125, 127, 129, 131, 133, 135, 137, 139–142, 145–147, 180
ergative 43, 52, 150–153, 158, 196, 204, 211–213, 228
exterior 111, 113, 125–128, 131–133, 135–142, 144–148, 180

Figure 50, 107, 115, 188, 197, 199

gender 35, 38, 43, 45, 53, 56–58, 66, 82, 84, 92, 93, 96–99, 200, 204, 208, 215
genitive 41, 54, 62–69, 75, 76, 78–82, 84–87, 92, 94–99, 104, 128, 137, 140, 150, 151, 155, 159, 160, 163, 164, 166, 168, 171–173, 177, 178, 188, 211, 219, 225
– inflectional genitive 62, 69, 70, 75, 76, 79, 80, 83–85, 96–99, 103, 104
grammaticalisation/grammaticalised 115, 130, 183, 188–190, 200–207
Ground 5, 6, 11, 30, 36, 40, 41, 50, 107, 115, 145, 160, 162, 163, 178, 180, 183–189, 191–193, 196–199, 208–211, 218, 219, 221, 228

heterogeneity/heterogeneous 15, 70, 85, 99, 165, 181
homogeneity/homogeneous 2, 14, 28, 31, 105, 124, 133, 148, 171, 181, 196, 224

illative 111–113, 115, 118, 121, 122, 125–127, 129–131, 133–135, 138, 139, 141–147, 180
inessive 108–115, 118, 119, 121, 122, 125–127, 131, 133, 135, 137, 138, 141, 142, 145, 180
inflection/inflectional 16, 19, 30, 38–41, 46, 51–54, 58, 62, 66, 67, 69, 70, 75, 76, 78–80, 83–85, 93, 94, 96–99, 103–109, 111, 112, 125, 126, 129, 133, 135, 137, 138, 140, 156, 159, 160, 171, 174, 176, 178, 181, 182, 196, 225
instrumental 150, 151, 157, 178
interior 110–113, 115–122, 124–128, 131–133, 135–142, 144–148, 180

lative 41, 115
location/locational 4, 10, 13, 28, 32, 40, 59, 100, 106, 108, 109, 115, 128, 163, 164, 179, 192, 200, 207, 209, 210, 220
– general location 40, 106, 108, 109, 112, 114–118, 122–124, 126–132, 135–139, 141, 142, 147–149, 170, 181, 191, 221
locative 4, 13, 29–31, 35, 39–41, 106, 110, 113–115, 118, 119, 122–124, 132, 150–152, 154, 156, 157, 159–161, 163, 172, 173, 176–178, 180, 183–188, 194, 195, 198, 207, 208, 218, 224, 225, 228

mismatch 15, 16, 19, 22, 24, 30, 40, 52, 58, 81, 99, 105, 109, 165, 181, 197, 224–227
motion 5, 11, 40, 75, 107, 140, 170, 186, 188–190, 192, 210
morphophonology/morphophonological 119, 134, 154–156, 166
morphosyntax/morphosyntactic 1, 3, 18–20, 22, 23, 31, 32, 34, 36, 38, 41, 43, 46, 48–51, 53, 55, 58, 92, 100–102, 104, 107, 168–170, 175, 176, 178, 179, 181, 182, 190, 192–195, 199, 205, 207, 209, 224, 225, 227, 229

nominative 53, 57, 65, 75, 78, 79, 81, 159, 160, 171, 172, 228
number 1, 15, 16, 28–31, 43, 52, 57, 58, 98, 123, 131–133, 150, 151, 159

overabundance 118, 119, 123, 132, 135, 142, 144, 153–155, 165, 225
overdifferentiation 16, 52, 135, 225

partitive 150, 151, 164
periphrasis 109, 116, 181, 225
– anti-periphrasis 40, 181, 225
perlative 39, 40
postposition 54, 61, 81, 108, 110, 115–118, 128, 129, 151
– postpositional phrase 61, 108, 109, 115, 117, 128, 130, 151
possession 9, 20, 29, 60, 61, 65, 66, 76, 78, 81–83, 85, 86, 88, 91, 100–104, 210, 211, 219, 220, 225, 227
– adnominal possession 49, 59–61, 65, 69, 75, 81–84, 86, 91, 100, 101, 103, 221,
– alienable possession 83, 90, 211
– inalienable possession 100, 101, 210, 211, 222, 223
– predicative possession 101, 125, 129, 138, 139
possessive 57, 60, 61, 64, 66, 70, 75, 76, 78, 82, 84–86, 88, 95, 100–104, 151, 211, 221, 222
– possessive construction 30, 53, 69, 70–73, 75, 82–91, 104, 222, 223, 228
– possessive relation 61, 63, 73, 76, 101, 138, 222, 223, 227
possessee 30, 59, 61, 63, 66, 67, 70, 72, 73, 75, 76, 79–87, 89, 90, 92, 95, 98, 100, 101, 103, 210, 219, 222, 223

possessor 9, 11, 30, 49, 59, 60–84, 88–92, 95–104, 125, 129, 138, 139, 210, 211, 219–223, 225, 228
– ANTH-possessor 60, 63, 71–73, 76, 82, 84, 90, 91, 97, 100, 103, 104, 210, 211, 219, 220, 222, 223, 227
– COMM-possessor 66, 71, 76, 100, 103, 104, 210, 211, 219, 225, 227
preposition/prepositional 7, 33, 34, 36, 40, 46, 54, 66, 67, 69, 70, 75, 76, 78–81, 83–87, 89, 90, 93, 159–164, 169–181, 183, 185–194, 205, 208–210, 217, 218, 220, 221
pronominal 72, 75, 76, 96, 192
pronoun 24, 25, 53, 54, 57, 61, 65, 67, 69–71, 78, 82–84, 88, 96, 100, 102–104, 134, 165, 184, 192, 197, 202, 215, 221
prototypical/prototypicality 9, 18, 24–26, 29, 36, 44, 46, 49, 50, 100–104, 149, 210, 213

relative 30, 33–35, 150–158, 184

semantic/semantics 17, 24, 36, 37, 41, 54, 56, 75, 91, 92, 99, 109, 115, 137, 145, 148, 149, 156, 163, 168, 190, 204, 206, 207, 212, 220
spatial case 38, 108, 110, 115, 116, 120–128, 130–133, 135–139, 141, 142, 144, 147–153, 156–158, 198, 208, 229
spatial relation 20, 30, 35–42, 46, 106, 107, 109, 110, 112–114, 117, 118, 122, 126, 127, 129, 140, 142, 144, 146, 159–161, 163–168, 171, 173–180, 182, 183, 186, 187, 190–196, 198, 199, 207–210, 217, 219, 227, 229
static 108, 114, 115, 126, 128, 142, 160, 165, 183, 185, 186, 190–192
sublative 111, 112, 115, 116, 121, 122
suffix/suffixal 11, 29, 30, 35, 39, 41, 45, 51, 52, 54–57, 66, 78, 96, 98, 99, 104, 108, 112, 113, 117, 118, 122, 150, 151, 154–158, 164, 183, 185, 196, 205, 214, 218, 219, 222
superessive 108–119, 121, 122
superior/superiority 108, 110–113, 115–122, 124, 125, 128
syntax/syntactic 1, 2, 5, 15, 19, 23, 25, 34, 36, 38, 43, 56, 66, 72, 73, 76, 91, 95, 98, 99, 106, 118, 136, 149, 163, 168, 175, 177, 196, 198, 205, 209, 211, 212, 216, 225

terminative 111, 112, 133, 150–156

universal 19, 32, 43, 228

vocative 3, 14, 44–46, 54–58, 64, 159, 171, 201, 203, 230

zero-marking/zero-marked 25, 37–39, 41, 46, 61, 62, 103–106, 135, 152, 185, 187, 189, 194, 195, 198, 199, 207–211, 213, 214, 216, 218, 222–227, 229

www.ingramcontent.com/pod-product-compliance
Lightning Source LLC
Chambersburg PA
CBHW050519170426
43201CB00013B/2018